Demonstrating Science with Soap Films

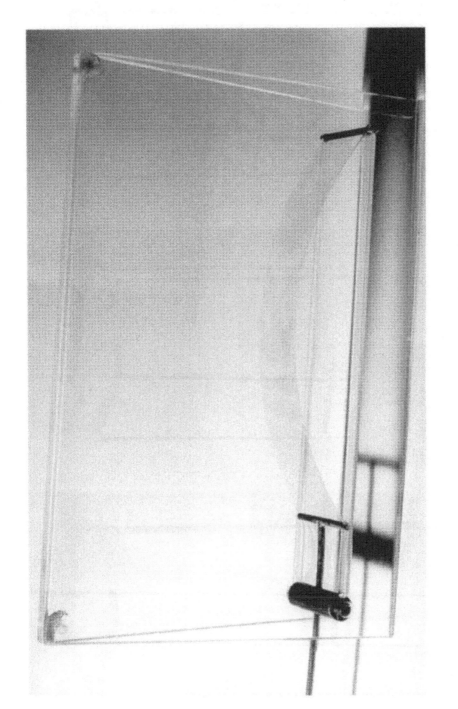

Frontispiece F1. Soap film within a wedge.

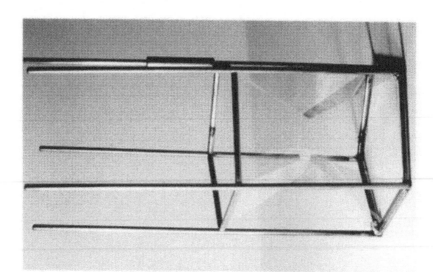

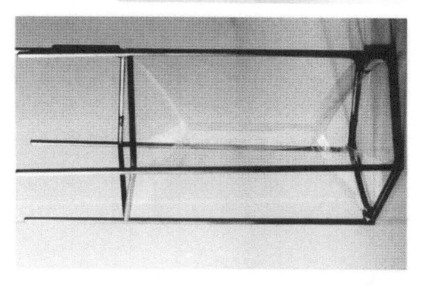

Frontispiece F2. Photographs showing the jump between the two soap-film configurations within a framework shaped as a pentagonal prism.

Demonstrating Science with Soap Films

David Lovett

Department of Physics
University of Essex

Computer Programs

John Tilley and David Lovett

CRC Press
Taylor & Francis Group
Boca Raton London New York

CRC Press is an imprint of the
Taylor & Francis Group, an **informa** business

CRC Press
Taylor & Francis Group
6000 Broken Sound Parkway NW, Suite 300
Boca Raton, FL 33487-2742

First issued in paperback 2019

Typeset in TEX by Institute of Physics Publishing

No claim to original U.S. Government works

ISBN-13: 978-0-7503-0269-2 (hbk)
ISBN-13: 978-0-367-40213-6 (pbk)

British Library Cataloguing-in-Publication Data

A catalogue record for this book is available from the British Library.

Library of Congress Cataloging-in-Publication Data

Lovett, D. R.
   Demonstrating science with soap films / David R. Lovett ; computer
programmes, John Tilley and David Lovett.
      p.   cm.
   Includes bibliographical references and index.
   ISBN 0-7503-0270-4. -- ISBN 0-7503-0269-0 (pbk.)
   1. Soap-bubbles. 2. Soap-bubbles--Study and teaching.
I. Tilley, John, Ph. D.   II. Title.
QC183.L85 1994
530.4′275--dc20
                                                                94-30453
                                                                CIP

Visit the Taylor & Francis Web site at
http://www.taylorandfrancis.com

and the CRC Press Web site at
http://www.crcpress.com

Contents

Preface

For many, the blowing of soap bubbles as children formed a significant part of their introduction to science. By generating these bubbles, children meet ideas of shapes and forms which can be of considerable surprise even within very simple situations. They encounter the idea of stability, or perhaps more importantly the idea of instability as the bubbles burst. And they see something of the beauty which science can produce as they watch the bubbles catch the light to produce startling colours as a result of the interference of light within the surfaces of the bubbles.

Throughout scientific history, soap films have held the interest of scientists and mathematicians with an equal intensity to that of children. A topic which, on the one hand, is very simple with few variables and constraints proves, on the other hand, to be difficult to solve in its entirety by mathematics and to embrace fundamental principles which underlie all of physical science.

This book attempts to pass on some of the fun which can be obtained from studying soap films and soap bubbles. At the same time, it has a much more serious objective, that of using films to help one understand principles of physics in particular and other aspects of science in general. This is possible because soap films are forever attempting to minimize their energy. In the same way, minimization of energy is fundamental to the setting up of equilibrium physical situations. So this book is about soap films; but it is also about the importance of *energy minimization* to the establishment of the world around us. And it is about the *patterns* which are set up as a result of this minimization.

In most topics in physical science it is difficult to avoid mathematics. The study of soap films is no exception and simple calculations and algebraic equations are dispersed throughout the book. Where the amount of mathematics is more significant, the relevant section has been boxed and can be left out if desired.

Although the balance of content reflects very much my own emphases and interests, the development of the material has arisen from considerable interaction with colleagues and friends. I first became interested in the topic after attending, in close

The author demonstrating the wonder of soap films to a captivated audience. Picture by courtesy of Essex County Newspapers Ltd.

succession, public lectures by Dr Cyril Isenberg on soap films and soap bubbles and by Professor Christopher Zeeman on catastrophe theory. Following this, the interest of colleagues in the Department of Physics here at Essex University has been a great encouragement. In particular, I must acknowledge the considerable input from Dr John Tilley. Not only has he collaborated with me over many of the ideas, but he is making a major contribution to one of the appendices, and has provided numerous helpful suggestions concerning the overall text. Over the years, Dr Stephen Smith, Professor Rodney Loudon and Professor David Tilley have made many helpful suggestions. I should thank Brian Diamond and John Bartington who made the models, modified them with considerable originality, and sometimes repaired them. The staff at Institute of Physics Publishing have been most helpful in the preparation of the text. I am indebted to Jim Revill for his enthusiastic encouragement of the project and to Al Troyano for his contribution to the layout and his efficient transformation of the manuscript into a most attractive book.

Above all, this book is a result of the fun and excitement I have had with hundreds of enthusiasts who have marvelled at the amazing patterns set up by soap films.

David Lovett
University of Essex
December 1993

1

Introduction

The nature of soap films and films produced by other surfactants is described. The importance of surface energy is explained, and minimal-area surfaces are introduced. It is the comparison of minimization of energy and other quantities in physics with the minimization of area of a soap film which underlies all subsequent topics in this book.

1.1 WHAT ARE SOAP FILMS?

To create a soap film we must first produce a soap solution consisting of soap molecules and water molecules. Soap molecules consist of sodium stearate (or some other fatty acid salt), although the term soap film is now a generic term for a wide range of other molecules which produce similar effects. Sodium stearate is a long-chain fatty acid of composition $C_{17}H_{35}COONa$. Immersed in water, the molecules become ionized. The sodium ions have positive charge and disperse throughout the solution. This leaves a negatively charged head to the stearate ion which has a long hydrocarbon tail. There is a nett force on the stearate ions tending to take them to the surface to leave the hydrocarbon tails sticking out (figure 1.1(a)). Because these tails ($C_{17}H_{35}$) stick out of the water, they are often described as hydrophobic or water-hating and the so-called carboxyl head (COO-) as hydrophilic or water-loving. Some of the stearate ions remain within the bulk solution and play no part in the film. The same arguments apply if one uses artificial soaps and in particular the synthetic detergents as used for household washing-up liquids. Such molecules are often called surfactants because of the way they establish themselves at the surface.

Given sufficient encouragement via the processes of blowing bubbles or dipping frameworks into them, soap solutions can be induced to establish films in which two parallel surfaces of surfactant molecules are set up with water solution between (figure 1.1(b)). The surfactant ions have a length of approximately 3 nm (30 Å) and an area of 0.40 nm² (40 Å²). The actual films can have a wide range of thicknesses varying from approximately 5 nm (50 Å) to ~ 10⁴ nm (10⁵ Å) according to the volume of solution between the surfactants. We shall return to the topic of film thickness and the corresponding interference colours later in the book (see page 77).

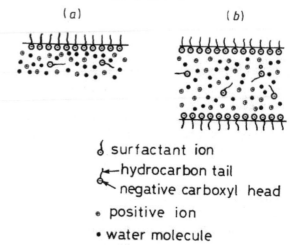

$\oint$ surfactant ion

$\longleftarrow$ hydrocarbon tail

$\searrow$ negative carboxyl head

• positive ion

• water molecule

Figure 1.1 (a) Structure at the surface of a soap solution. (b) Structure through a soap film.

The lifetime of pure soap films is very sensitive to impurities such as dust particles and components such as excess fat. This does not apply in the case of modern detergents. Using tap-water plus 1 to 2 per cent washing-up liquid (liquid detergent), such as Sunlight (in the UK) or Dawn and Joy (in the USA)†, will produce films which last for at least 15 seconds. They can last considerably longer within some of the frameworks to be discussed later. Higher concentrations of washing-up liquid, say 10 per cent concentration, should enhance the lifetime of the films further. A moist atmosphere increases lifetime, whereas films tend to be short-lived when demonstrated above an overhead projector. Adding glycerol increases the lifetime of the films to times of minutes with say 5 per cent glycerol and to hours for 50 per cent glycerol. However, adding glycerol, if not adequately mixed in, can produce problems of cleaning

† Other examples in the UK include Persil washing-up liquid, Squezy and many 'own-brand products' available from superstores. The suitability of these and other washing-up liquids may change as their compositions are varied.

the frameworks after use, whereas the lifetime achieved with commercial liquid detergent is usually adequate. It is very important that the detergent is thoroughly stirred into the water and does not form a residue on the bottom of the container. Leaving the solution for a time prior to use is helpful.

Recipes have been given for solutions which produce longer-lasting films and in particular Stong (1969) has described a number of alternative solutions which can be used. A solution first described by Plateau (1873) can be produced as follows. To 1200 g of distilled water add 30 g of sodium oleate. The latter which should be chemically pure is a white powder that floats on the surface of the water. Rather than mixing the powder in with the water, it is advised to allow the water plus powder to stand for 24 hours by which time the powder will have dissolved. At this stage, 30 g of glycerol is added and the solution mixed by pouring it back and forth between two containers. The mixture is stored in darkness for a week after which time the clear fluid which is formed below a scum on the top is syphoned off. A few drops of ammonia are added and the solution stored for use. Bubbles from this solution last for minutes.

A recipe reported by Jearl Walker (1987a) in *Scientific American* for producing rigid films is as follows:

> 100 g glycerine (85% solution)
> 1.4 g triethanolamine
> 2.0 g oleic acid

The chemicals are mixed and stored for 24 hours in the dark in an airtight bottle. If the solution is not clear at this stage, a little more triethanolamine should be added.

Kuehner (1958) described the preparation of solutions which enable bubbles to last for years. His method is very detailed and requires control of temperature often to below -20 °C. First oleic acid is purified. Secondly the purified oleic acid is converted to 9,10-dibromostearic acid by the addition of bromine. Thirdly, sodium 9,10-dibromostearate solution is produced by carefully neutralizing the acid with hydrogen dioxide and mixing-in glycerol. This solution produces long-lasting bubbles. This method of Kuehner, and a further method of Stong in which two solutions are mixed, one being Kuehner's solution and the other being a solution of polyvinyl alcohol, water and glycerol, are not advised for the home experimenter. Rather the methods are mentioned to illustrate that for the experiments described in this book the use of selected commercial washing-up liquids is both adequate and very much easier.

1.2 SURFACE TENSION AND SURFACE ENERGY

Molecules near to the surface of a pure liquid do not exist in a uniform environment as do those within the interior of the liquid. They experience a weaker force in the direction of the gaseous region outside the liquid than in a direction towards the bulk liquid. Hence the molecules experience a nett force pulling them inwards and this has the effect of trying to reduce the volume of the liquid. This inward force is opposed by the outward forces between the molecules within the interior of the liquid as the molecular separation infinitesimally decreases. In addition, the resulting density of the liquid is slightly less at the surface. Thus we can see why water droplets take up a spherical shape. There is a good analogy with the situation of a balloon. The stretched rubber of the balloon attempts to pull the outer surface inwards and the pressure within the balloon counteracts this effect.

For the surface of the liquid to be in a state of uniform surface tension, the surface tension must be perpendicular to any line drawn in the surface. It must also have the same magnitude for all directions of the line within the surface, and it must have the same value at all points over the surface. In the case of a film, there are two parallel surfaces and hence an overall force of double magnitude.

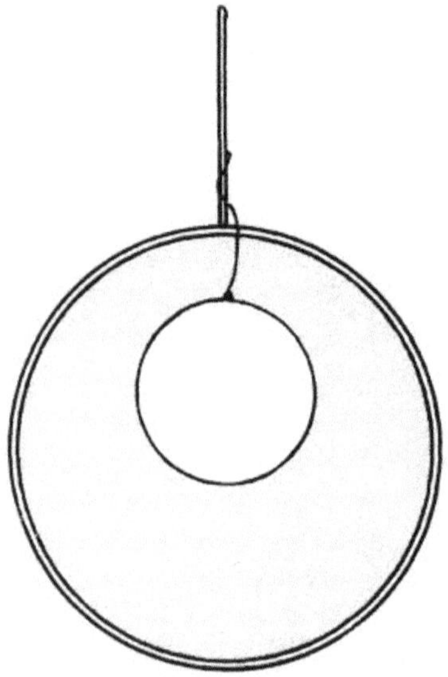

The consequence of the above can be shown rather nicely using a wire hoop and a length of thread. The hoop is dipped into soap solution and removed to produce a uniform film. A loop of the thread is placed on the film and the film punctured within the loop. As a result the thread is pulled outwards by the film to form a very precise circle (figure 1.2). Now it is a property of a circle that it possesses maximum area for its given perimeter. More importantly, the film outside the thread takes on minimum area and as we shall soon realize this minimizes the energy associated with the film.

The above demonstration gives us the hint as to how we might define and measure the surface tension γ. We shall define γ as the force per unit length acting perpendicular to one side of a line within the surface of a liquid; but so that we can measure its value we shall need to have a film on one side of the line and air on the other. A suitable method of measuring the value is to use a sliding wire moving across a fixed frame (figure 1.3). We apply a force F to the right of the wire to balance the force due to surface tension on the left. At balance, the force arising from the surface tension will be $2\gamma l$, where l is the length of

Figure 1.2. Demonstration of minimization of film area.

the sliding wire and the factor of 2 arises because there are two film surfaces. The surface tension of water is approximately 73×10^{-3} N m^{-1}, but for soap solutions or oil or methylated spirit the values are considerably less. Hence, if one drops a little of any of these other liquids onto the surface of the water, assuming that the water surface is very large, the additive spreads out to form a monolayer.

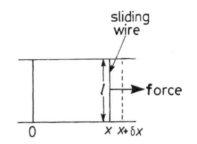

Figure 1.3. Framework to measure surface tension.

The arrangement of figure 1.3 allows us to obtain an expression for the surface energy of the film. Suppose we do work using a force F (strictly we use a force infinitesimally greater that F) to move the wire from position x to position $x + \delta x$. The work done will be $2\gamma l\delta x = 2\gamma \delta A$ where δA is the change in area of the film. The total energy to produce a film of area A, starting from zero area, will be

$$E = \int_0^A 2\gamma \, \mathrm{d}A = 2\gamma A$$

where E is the free energy of the film at constant temperature. The important feature is that the free energy of the film is proportional to its area as is demonstrated by the equation. Soap films attempt to minimize their energy and hence this means that they attempt to minimize their area. This minimization is fundamental to what follows in this book and underlies a large number of the demonstrations and analogues which can be shown using soap films.

1.3 A BRIEF HISTORY OF SOAP-FILM STUDIES

The understanding of the formation of various soap-film patterns is intimately linked with the mathematics of the calculus of variations and various principles of minimization. Many well-known mathematicians have used soap films to help them to understand or demonstrate their mathematical ideas. So perhaps we should start with Maupertuis (1698–1759) who stated his general principle that

'if there occurs some change in nature, the amount of action necessary for this change must be as small as possible'.

The minimization of the energy of a soap film is another aspect of the theorem and it is the general nature of the theorem originally proposed by Maupertuis that allows soap films to be used for such extensive illustrations of other aspects of physics. *Maupertuis' Principle of Least Action*, the idea that nature always minimizes action, is another form of his general principle. However, it is perhaps surprising that a further application of Maupertuis' Principle, the so-called *Fermat's Principle*, preceded the work of Maupertuis and its publication. (Pierre de Fermat, a French mathematician, lived 1601–1665.) The principle states that light passes as rays along a path between source and observer such that the time needed is less than the time required for any alternative path. If we consider light passing through a single homogeneous medium, Fermat's Principle reduces to the simplified form of stating that the light path between source and observer is the shortest path geometrically, a fact know even to the Greeks. Prior to Maupertuis stating his principle, Euler (1707–1783) had already written the first book on calculus. He had expounded the idea that behind phenomena of the universe in general there is always a rule involving maximization or minimization.

Although many of the phenomena associated with soap films were known prior to the second half of the nineteenth century, the most extensive early practical work was carried out by the Belgian physicist Joseph Plateau (1802–1883), who wrote a two-volume treatise on the subject in French (Plateau 1873). Plateau's contribution to the subject was especially remarkable in that he had become almost blind from observing the sun for 25 seconds during an experiment in 1829. In his treatise on soap films he describes their formation within three-dimensional frameworks, in addition to describing practical aspects of establishing the films and outlining some of the underlying mathematics. The nature of his work is reflected in the translated title of the book, *Experimental and Theoretical Investigations of Equilibrium Properties of Liquids Resulting from their Molecular Forces*.

Josiah Willard Gibbs (1839–1903), well-known for his contributions to thermodynamics and statistical mechanics, investigated the draining and thinning of soap films. But the best known name in the field is that of Sir Charles Vernon Boys (1855–1944) who gave lecture-demonstrations on the subject and wrote these up in a book (1890, reprinted 1959), the more recent title of the book being *Soap Bubbles, Their Colours and the Forces which Mould Them*. It has remained in print and is an excellent short read. Sir James Dewar (1842–1923) also researched soap films, and the results of his work are best described in the popular text by his assistant A S C Lawrence who wrote *Soap Films, a Study of Molecular Individuality*.

Following on within this tradition in the United Kingdom, Cyril Isenberg has given an extensive number of lecture-demonstrations, showing both soap patterns in two-dimensional and three-dimensional frameworks and illustrating the interference patterns observed in such films. These aspects are described in *The Science of Soap Films and Soap Bubbles* (1978, 1992). In the USA, Frederick J Almgren and Jean E Taylor have popularized their work on 'The geometry of soap films and soap bubbles' in *Scientific American* (1976), and the geometrical and mathematical aspects have been taken further in the *Scientific American* book *Mathematics and Optimal Form* by Stefan Hildebrandt and Anthony Tromba (1985).

Further main contributions have tended to concern the mathematics of minimal surfaces and in particular Plateau's problem which will be introduced in the following section of this chapter. J C C Nitsche has discussed the mathematics of minimal surfaces extensively in *Vorlesungen über Minimal Fläschen* (1975), part of which has been translated into English as *Lectures on Minimal Surfaces*, volume 1 (1989).

The particular aspect of soap films on which this book will concentrate is that of the changes of film shape which occur when the frameworks alter their sizes. Discussion of these changes will involve, in addition to the mathematics of minimization, the more recent mathematics of catastrophe theory and phase transitions. These other topics will be introduced in a later chapter.

1.4 PLATEAU'S PROBLEM

In carrying out his experiments, Plateau realized that any single closed wire, whatever its shape, can bound at least one soap film. This is, of course, assuming that the closed wire is not so large that the film cannot maintain its own mass. He went on to ask the question as to whether every closed contour in space (i.e. closed wire) can be spanned by at least one minimal surface. Empirically, this is what soap films appear to show. However, mathematicians have found it a very intriguing and difficult problem, except in simple cases.

A shape which mathematicians like to use as a starting example is that shown in figure 1.4(a). This framework can be spanned by at least three minimal surfaces as shown in figures 1.4(b), (c) and (d). Although the surfaces shown are all minimal surfaces joining the wire framework at all points, they are likely to have three different magnitudes of area, the precise magnitudes

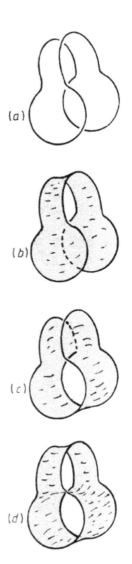

Figure 1.4. Wire framework shaped like headphones (a) and the three possible minimal surfaces which can form within the framework (b) to (d).

depending on the actual curvatures along different sections of the wire framework, according to just how the wire has been shaped. Only one film surface will be the absolute minimum of area to join the entire framework. In addition, and very importantly for mathematicians in their analytical approach, the surfaces shown in figures 1.4(b) and (c) are of one mathematical type, whereas that in figure 1.4(d) is of a different type. Mathematicians refer to the genus of a surface in order to differentiate types. A surface has genus g if it is equivalent to a sphere with g handles attached to it. Thus a lidded teapot with one handle has genus 1 and a lidded sugar bowl with two handles has genus 2. Any surface which is a distortion of a sphere, including the distortion of a sphere which starts with one or more holes cut from it, will be of genus 0. A disc will have genus 0 and the films in figures 1.4(b) and (c) are of genus 0. A torus (i.e. a doughnut) has genus 1 and the film in figure 1.4(d) also has genus 1. The wire framework in figure 1.4 is very simple. It demonstrates that even deciding how many minimal surfaces can exist bounded by a given contour can be a difficult problem.

It has now been proved that for each simple closed curve, there exists a surface of least area and of finite genus spanning this contour; in addition, this least-area surface will not have self-intersections. The last part of the statement means that if curves intersect (are knotted), the least area cannot be a disc (i.e. bounded by a simple closed curve). However, there are difficulties in knowing the number of minimal surfaces, or even the number of disc-type minimal surfaces that span a non-planar contour.

> There is no example of a curve that bounds more than one minimal surface and for which *all* minimal surfaces that span it (even of a specific genus) are known, nor are there even any examples that indicate plausible answers.
> *Hildebrandt and Tromba* (1985, p 110)

It is no wonder that the minimal surfaces exhibited by soap films are so interesting.

From his experimental work, Plateau formulated two rules:

Rule 1. If a minimal surface has a free boundary on a surface of support, then it meets that boundary at right angles.

In this book we shall be looking at the films established between perspex plates, metal (or other) pins, and metal (or other) frameworks. The films will always meet the perspex plates at right angles and will always meet the cylindrical wires at right angles to the tangent to the surface at that point.

Rule 2. In three dimensions, three smooth minimal surfaces, of an area-minimizing system of surfaces, intersect in a smooth line at an angle of 120° to each other. Only four such lines, each formed by the intersection of three surfaces, can intersect at a single point. At such a point, the angle between adjacent lines is 109°28'16".

The angle 109°28'16" may appear to be an arbitrary angle but is the angle whose cosine is $-\frac{1}{3}$. One approach to proving the angle is to look at the geometry of a soap film within a tetrahedron (see page 99, figure 6.6). Four film components meet centrally and there are four symmetrically arranged lines of intersection meeting up at this central point. These lines of intersection are the same as the bonding directions for the tetrahedral bonds found in so many covalent crystals including diamond, germanium and silicon.

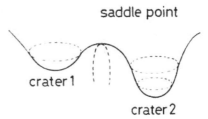

Figure 1.5. Terrain involving two craters and a saddle point.

1.5   THE SHAPE OF MINIMAL SURFACES

A minimal surface will have a minimum value of free energy although this minimum may be a local minimum only. By analogy, if we sit at the bottom of a crater of say an extinct volcano, we can descend no lower in that region, but we might be able to climb up over the lip and into a second crater which descends further (figure 1.5). In both positions, which are at the bottoms of the craters, we are in stable equilibrium. But the bottom of crater 2 corresponds to a lower potential energy than that of crater 1. Our figure illustrates a further complication that can similarly arise with soap films. In walking up the side of one crater and down the side of the other we can pass over the lip through a position of unstable equilibrium. To pass over this precise position of unstable equilibrium, we must walk over the so-called saddle point. In fact, to have two minimal soap-film surfaces (of disc-type) of stable equilibrium requires the existence of a third surface (also of disc-type) corresponding to unstable equilibrium. In flipping from one to the other stable situation, we pass swiftly through the third, unstable, surface.

We now ask what will characterize an equilibrium surface, whether existing in stable or unstable equilibrium. The answer lies in the curvature of the surface or film. If we have a circle of radius R, we define the curvature κ as being equal to the inverse of the radius, i.e.

$$\kappa = \frac{1}{R}.$$

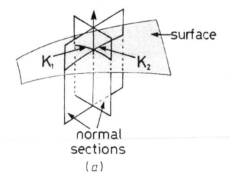

normal
sections

(a)

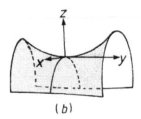

(b)

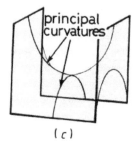

principal
curvatures

(c)

Figure 1.1. (a) Illustration of principal curvatures for a surface. (b) Principal curvatures for a saddle point. (c) Illustration of a saddle point by a soap film joining a simple wire framework.

Hence large circles have small curvature, and *vice versa*. This fits with our concept of a straight line having no curvature and being equivalent to a circle of infinite radius.

This situation of a single curvature is fine for a circle in two dimensions, but when we have a surface within three dimensions, such as a saddle surface, we shall have a number of curvatures whose values vary according to which cross-section we consider. What we require are the two principal curvatures, κ_1 and κ_2. These are the maximum and minimum curvatures which can exist within a normal section through the point of interest on the surface (figure 1.6(a)). For a saddle point it is easy to see the sections we should use; these are the xz and yz planes in figure 1.6(b). The principal curvatures are in orthogonal planes (i.e. planes at right angles to each other). This orthogonality is always so for the two principal curvatures, whatever the small piece of surface we consider. Commonly, the direction of one principal curvature is obvious, and then we can find the direction of the other. We may then *define* a Gaussian curvature κ as

$$\kappa = \kappa_1 \kappa_2$$

and a mean curvature H as

$$H = \frac{\kappa_1 + \kappa_2}{2}$$

A surface of minimal area must have a mean curvature which is zero, i.e.

$$H = 0 \qquad \text{or} \qquad \kappa_1 = -\kappa_2.$$

The physical explanation of this lies in the Laplace–Young equation which relates excess pressure P across a surface at any point to the radii of curvature at that point;

$$P = \gamma \left(\frac{1}{R_1} + \frac{1}{R_2} \right)$$

where R_1 and R_2 are the maximum and minimum radii of curvature (see for instance Newman and Searle 1957, p 166). The equation can be obtained by considering the excess pressure acting on the concave surface of a curved membrane under uniform tension. Hence for no excess pressure, H is zero. Note that for a spherical bubble of radius R with two surfaces, this equation gives us

$$P = \frac{4\gamma}{R} \qquad \text{or} \qquad P = 4\gamma H$$

so that with an excess pressure inside a bubble, there must be finite curvature.

Similarly, surfaces corresponding to *unstable* equilibrium must have a mean curvature of zero. Sometimes these surfaces are also referred to as minimal surfaces. All minimal surfaces, including this group, are either flat or look like saddle surfaces. This feature therefore characterizes all the films which we shall consider in this book, except where we include bubbles with an excess pressure.

1.6   THE SADDLE-POINT BIFURCATION

We illustrate in figure 1.6(c) a saddle point established by a soap film within a wire framework. The components of the framework, all linear, correspond to some but not all the edges of a cube. This model provides a simple and very clear way of demonstrating a saddle point. If we construct the framework to be longer in one direction as shown in figure 1.7, then the saddle point can exist in either of two positions within the framework. By suitably distorting the frame, the saddle point can be caused to switch from one position to the other (Mackay 1985b). The switch can be produced either by opening out or closing up pairs of the longer sides at one (or other) end of the model. Because the saddle point can take up either of two non-central positions, the phenomenon is called a bifurcation.

We shall see that the sudden switch between positions is analogous to a change of phase as studied in thermodynamics and as demonstrated in crystal structures. In the case we are looking at here, the symmetry or pattern of the film does not change between alternative configurations; it merely rotates and reflects. Sometimes this is what happens in crystals. Further, we shall discover that such a change fits the description of a catastrophe in a branch of mathematics called catastrophe theory.

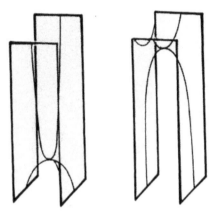

Figure 1.7.   Bifurcation of a soap-film saddle point.

2

Two-dimensional soap-film patterns

Soap films are constrained between parallel plates such that their overall length is minimized in order to minimize energy. The equilibrium film patterns established to link up various arrangements of pins are described. Within these patterns, wherever film components meet, they do so in threes, always at an angle of 120° to each other. As a consequence of this condition, interesting geometrical relationships arise.

2.1 INTRODUCTION

As discussed in Chapter 1, soap films attempt to minimize their area in order to minimize their energy; or alternatively they readjust their shape to produce a local minimum of area such as to minimize their energy *locally* (subject to a small perturbation of the film). If we confine the films between two parallel plates then minimization of the area is equivalent to minimization of the length. These parallel plates are most suitably made from perspex. We then insert pins across the gap between the two plates. These pins can be fixed into the perspex, although simpler arrangements can be used, such as resting the top plate onto the points of drawing pins, which themselves are resting on the bottom plate. For further discussion on how to make the models see Appendix I. The problem of establishing what happens to the film becomes a two-dimensional one of seeing how the film links up the pins. It is identical to the problem in two dimensions of the minimum length required to join a number of points.

2.2 SIMPLE TWO-DIMENSIONAL PATTERNS

2.2.1 Films joining triangular, square and pentagonal patterns of pins

It is clear that the minimum length to join two pins is the straight line between these two pins. The pattern to join three pins arranged at the vertices of an equilateral triangle is also obvious. Three arms of the film meet at the circumcentre of the equilateral triangle. The angle between each arm of the film is precisely 120° and the three component film surfaces are of equal lengths. Though one fully expects this pattern, nevertheless it comes as something of a surprise to see the film arranging itself with total precision to form the three equal arms (see figure 2.1). Assuming each side of the equilateral triangle is of length a, the length of each of the three soap-film arms equals $a/\sqrt{3}$. However, should the plates not be accurately parallel, then the film configuration is quite surprisingly different, and this will be considered later.

By the time we have four pins arranged at the corners of a square, the film configuration to join all four pins is quite unexpected. The film of shortest length does not pass along three sides of the square. This would produce a film of overall length $3a$, where a is the length of one side of the square. Nor does it form a cross joining opposite corners with the arms of the film intersecting at 90°. This arrangement would have a total length of $2\sqrt{2}a$ (i.e. 2.828a). Rather the film takes the shape of a distorted H and the angles between the separate parts of the film are all 120°. This is shown in figure 2.2. The total film length is $[a + \sqrt{3}a]$ (i.e. 2.732a) with the central 'bridge' of length $[a - a/\sqrt{3}]$ (i.e. 0.423a). The 120° pattern of angles between film components that meet is very general. The 120° configuration appears repeatedly within two-dimensional film patterns. We can easily understand this in terms of the film within the four-pin pattern. We can see that if it were possible for two films to cross at right angles, they would combine at the intersection, the single junction would open out into two separate junctions, and these would move backwards to form two 120° configurations, as already demonstrated at a single vertex within the three-pin configuration. This separation of the single junction into two and the subsequent realignment of the film components reduces the film length. At the same time the surface tension within the three film components balances out such that the film is in equilibrium.

What is interesting, and very significant for much that follows, is the fact that two equivalent soap-film patterns are possible for this four-pin configuration. These two alternatives are shown in projection in figure 2.2, one by continuous and one by dashed lines. Having established one of these configurations, it is relatively easy to blow the film such that it switches to the other equivalent configuration. To do this one holds the perspex

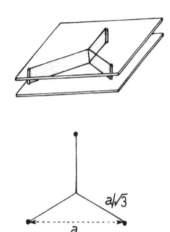

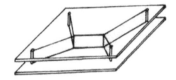

Figure 2.1. Two-dimensional film pattern established between three pins arranged in a triangle.

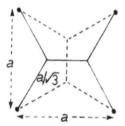

Figure 2.2. Two-dimensional film pattern established between four pins arranged in a square.

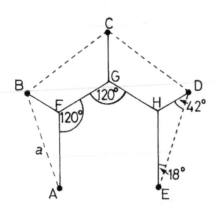

Figure 2.3. Two-dimensional film pattern established between five pins arranged in a pentagon.

plates horizontally and blows between the plates directly onto the film components to be moved.

Let us consider the pair of perspex plates with their associated four pins. Assuming that the pins are indistinguishable and that the perspex plates themselves are square, we cannot distinguish the setting of the plates in any one of four positions. Rotate the plates by 90° or by 180° or by 270° and nothing appears different. Yet once the film has been added, only the 180° rotation produces an equivalent position. We describe this change by saying that four-fold rotational symmetry has decreased to two-fold rotational symmetry. This is often referred to as 'broken symmetry'. But as we have seen, the two-fold configuration can be doubled by a film switch. Changes of symmetry in this manner can be encountered when one considers other examples of patterning such as in crystal structures. Also, the soap-film pattern has analogy with the structure of ethene (ethylene), C_2H_4, such that the positions of the hydrogen atoms correspond to the pins and the carbon atoms to the soap-film vertices. Ethene can flip between two configurations.

We can move onwards to five pins arranged symmetrically at the corners of a pentagon. The film of minimum length to join these pins takes the form shown in figure 2.3. This film has seven components to it, five arms linking directly with the pins, and two sections FG and GH which bridge between the arms of the film. We can see that the sections FG and GH could move round relative to the five pins, A, B, C, D and E, to a further four positions. Each of these arrangements would have identical energy assuming that the pins are precisely positioned, and physicists say that this minimum-energy arrangement has five-fold *degeneracy*. A pattern of pins with five-fold rotational symmetry has been turned into a pattern of pins-plus-film which has one-fold rotational symmetry, but there are five alternative degenerate arrangements. For a further discussion of rotational and mirror symmetries, see section 2.4.

2.2.2 Films joining a hexagonal pin pattern

Increasing the number of pins further, we consider a hexagonal pin arrangement. For the first time, the minimum length of film to join the pins is one which goes around the outside (figure 2.4(a)). The length will be 5a if the pin separation is a. We will number this as configuration I. Note that it has one-fold rotational symmetry but six alternative possibilities. If one also considers mirror symmetry with start and end points labelled, one then gets a further six indistinguishable possibilities. The fact that this arrangement (with its degeneracies) produces the minimum area of film does not mean, however, that we cannot link the pins with a film which cuts across the hexagonal area. The possibility of a film joining the pins by passing across the hexagonal area leads to two very distinct alternatives.

One alternative has three-fold rotational symmetry and is shown in figure 2.4(b). We will call it configuration II. Note that in this case pairs of adjacent pairs of pins are connected by sections of film, which either consist of two linking sections as in the case for the pairs of pins AB, CD and EF, or of four linking sections as is the case for pins BC, DE and FA. Rotation by 120° and by 240° produces indistinguishable patterns so the arrangement has three-fold degeneracy due to the three-fold rotational symmetry. A second arrangement for this configuration II can be obtained in which the entire pattern is rotated by 60° relative to the pins. It would not be easy, but in principle one should be able to blow the film across from one arrangement to the other. This new arrangement would look the same, but it would be distinguishable from the previous arrangement provided both the arms of the film and the pins are labelled separately. Once again one has three-fold rotational symmetry. So considering all the possibilities for this arrangement II, we find that the six-fold rotational symmetry of the plates-plus-pins has been converted to three-fold rotational symmetry; but there are the two equivalent arrangements and these lead to an overall degeneracy of 6. The length of film is the same in each case. Junction K is at the centre of the hexagon. The soap-film pattern is equivalent to that between three pins at the vertices of an equilateral triangle, but reproduced three times. An extra pin at K would not disturb the pattern. Hence, we can see that the overall film length is three times $\sqrt{3}a$, giving $5.196a$. We could also apply mirror symmetry to the pattern (increasing the overall total of distinguishable and non-distinguishable variations to twelve) but again no new patterning would arise.

The other configuration, number III, has two-fold rotational symmetry (figure 2.4(c)). We can deduce immediately the existence of, and then identify, three equivalent film arrangements as one rotates the film by 60° and then by a further 60° relative to the pins. Once again the film passes through the centre of the hexagon, so again a pin placed there would not disturb the arrangement. We can recognize that half of configuration III is equivalent to the film as established between four pins, although now these four pins are not placed

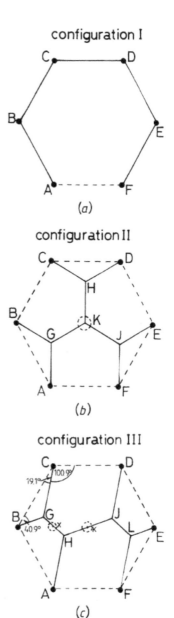

Figure 2.4. (a), (b) and (c): alternative two-dimensional film patterns which can exist between six pins arranged in a hexagon.

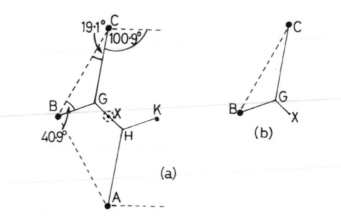

Figure 2.5 (a) and (b): steps in reducing the film pattern of configuration III within a hexagonal pin arrangement.

at the corners of a square (figure 2.5(a)). It is characteristic of complicated film patterns that they can be broken down in this way for analysis. In fact, this particular film pattern can be divided into two yet again (figure 2.5(b)). X is the intersection of the film GH with the line BE. This film pattern demonstrates some interesting classical geometry which will prove useful later in the book. This geometry is illustrated in the mathematical box overleaf (box 2.1). From the geometrical arguments demonstrated, the total film length for configuration III turns out to be $2\sqrt{7}a$ or $5.292a$.

We saw that in the case of configuration II various pattern and energy degeneracies were exhibited provided one assumes distinguishing labels for the arms and pins. There is an interesting and significant difference in the case of configuration III. There is doubling up of the number of distinct patterns if one takes into account mirror symmetry (equivalent to turning over the parallel plates). However, this brings the total overall number of distinguishable and indistinguishable possibilities once again to twelve.

This six-pin example is appropriate for considering the variation of total film length with a parameter which we will call the configuration parameter. What this parameter is precisely does not really matter. A starting suggestion might be to use the angle between a fixed direction and the direction of one arm of the film where it meets a pin. If we do this, we can immediately insert on a graph (see figure 2.6) the film lengths for the three configurations described. These are marked I, II and III and the

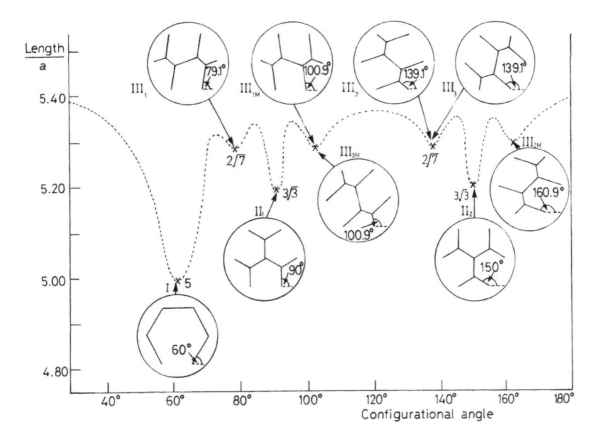

Figure 2.6 Plot of length (energy) as a function of a single configurational angle for the six-pin patterns described in figure 2.4.

numerical subscripts differentiate between the alternative angles possible for each of the configurations. Subscript M refers to a mirror version of a pattern. Hence the x axis of the graph has been labelled the configurational angle. To change from one minimum energy configuration to another would necessitate a major disturbance of the film by blowing the film vigorously. In making the change, the film must increase its length considerably before it relaxes back to the new arrangement. Hence we obtain peaks in the curve between the minima. These peaks have been sketched in on the graph but their precise form will depend on the precise nature of the perturbation set up. As the area of the film is proportional to the energy of the film, the parameter of length on the y axis of the graph can be written alternatively as a parameter of energy.

The minimum I corresponds to the film wrapping around the

Box 2.1   A useful geometrical equality

All the soap-film patterns are made up of repeated triple components of the film meeting at 120° to each other at a vertex. Let us consider such a vertex B with film components AB, BC and BX meeting at this vertex (see figure 2.7). Now draw the equilateral triangle ADC. Angles $A\widehat{D}C$, $A\widehat{C}D$ and $C\widehat{A}D$, being internal angles of the triangle, are all 60°. Angle $A\widehat{B}C$ is 120° as it is the angle between film components. $A\widehat{D}C$ and $A\widehat{B}C$ are opposite angles of the quadrilateral ABCD and are supplementary (i.e. their sum is 180°). This is the condition for a quadrilateral to be cyclic. Hence a circle can be drawn through the four points ABCD.

Let us call $A\widehat{C}B$ angle α. This angle must be equal to $A\widehat{D}B$ as they are angles within the circle subtended by the same chord AB. Similarly, $C\widehat{A}D$ equals $C\widehat{B}D$ equals 60° as both angles are subtended by CD; and $A\widehat{C}D$ equals $A\widehat{B}D$ equals 60° as both angles are subtended by AD. By using the sine equation for a triangle, and putting AC ($=$CD$=$AD) $= x$, we have the following:

$$\frac{DB}{\sin(60° + \alpha)} = \frac{x}{\sin(60°)} \quad \text{(triangle BCD)}$$

$$\frac{BC}{\sin(60° - \alpha)} = \frac{x}{\sin(60°)} \quad \text{(triangle BCD)}$$

$$\frac{AB}{\sin \alpha} = \frac{x}{\sin(60°)} \quad \text{(triangle ABD)}$$

$$DB - BC = \frac{x}{\sin(60°)}[\sin(60° + \alpha) - \sin(60° - \alpha)].$$

But $\sin(A + B) - \sin(A - B) = 2\cos A \sin B$; hence:

$$DB - BC = \frac{x}{\sin(60°)}[2\cos(60°)\sin \alpha]$$

$$= \frac{x \sin \alpha}{\sin(60°)} = AB.$$

Thus $AB + BC = DB$. This means that the total length of the soap-film components, AB, BC and XB, is equal to XD. This is an especially useful result in calculating film lengths within complicated patterns.

Box 2.1 *Continued*

It is easy to show that the radius of the circle ABCD is $x/\sqrt{3}$ (figure 2.8). We see this from calculating that length DY is $\sqrt{3}x/2$. Also $DZ = \frac{2}{3}DY$ (being a property of the circumcentre of an equilateral triangle). This gives us:

$$DZ = \frac{\sqrt{3}x}{2}\frac{2}{3} = \frac{x}{\sqrt{3}}.$$

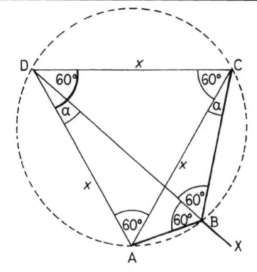

Figure 2.7 Calculating the total length of the film lengths meeting at a junction.

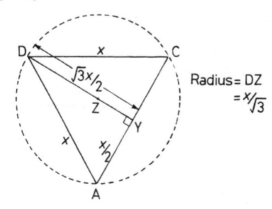

Figure 2.8 Calculating the radius of the circle used in figure 2.7.

outside of the hexagon and is the absolute minimum or global minimum. No other arrangement can be found which both joins the six pins and has shorter length or less energy. Minima II and III are local minima in that *major* displacement of the film can cause the film to jump to the lower energy state at I. *Small* displacements will only perturb the film locally from the specific type-II minimum or type-III minimum, and the film will relax back to its original type-II or type-III configuration once the displacement, or perturbation, is removed. Note that the positions marked I, II and III on the graph include degeneracy to cover all possibilities for the three configurations as described.

Unfortunately, we see that for two angular values, there are two configurational possibilities coinciding. They are both equivalent, merely involving a rotation of the film relative to the pins. However, they are distinct, requiring one to blow the film hard to rearrange it. This should tell us something. The patterning is two dimensional and one configuration parameter is inadequate. We need a second one. For the second parameter we use a second angle defined relative to the same direction as previously. We use the angular orientation of the film component linking up with a second pin adjacent to the first pin already used for defining the first parameter. We are now able to plot the variation of energy with respect to two configurational angles α and β, whereupon we can obtain an energy surface rather than an energy curve.

2.2.3 Films joining heptagonal and octagonal pin patterns

As we increase the number of pins, each pin set at a corner of a regular polygon, the tendency for the soap film to follow the perimeter becomes increasingly strong. It becomes quite difficult to blow the film into an alternative stable configuration linking pins via the internal area of the polygon. There is an additional complication. With these patterns of pins it is possible, in practice, to produce a number of apparently stable film arrangements, where the angles between components are slightly greater than 120°. This arises because some sticking of the film junctions on the perspex surfaces may occur. Such sticking may occur particularly if the junctions are near to the pins, which are of finite size. A departure from the ideal theoretical situation may occur, thus preventing the film moving to the correct pattern for a local minimum of energy. Stable patterns for heptagonal and octagonal pin arrangements are shown in figures 2.9 and 2.10 respectively.

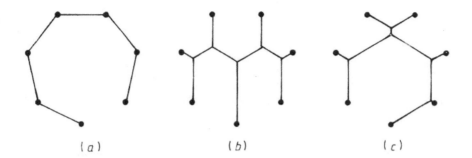

Figure 2.9 Alternative two-dimensional film patterns which can exist between seven pins arranged in a heptagon.

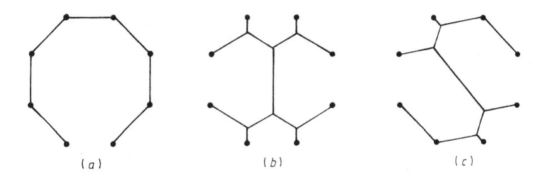

Figure 2.10 Alternative two-dimensional film patterns which can exist between eight pins arranged in an octagon.

2.3 BUBBLES WITHIN THE TWO-DIMENSIONAL PATTERNS

To trap a single bubble within a soap-film pattern, the perspex–pin arrangement is dipped into the soap solution to set up the film pattern. It is withdrawn from the solution and then reinserted to trap the required bubble. A little judicious blowing with a straw usually enables the removal of any undesirable complex patterning of the film, such as to leave a single bubble with as many sides as there are symmetrically arranged pins. Each corner of the bubble is linked by a linear film surface to one of the pins. The shapes of the bubbles formed are shown in figure 2.11. These shapes are predetermined by the need for the components of the film always to meet at angles of 120°.

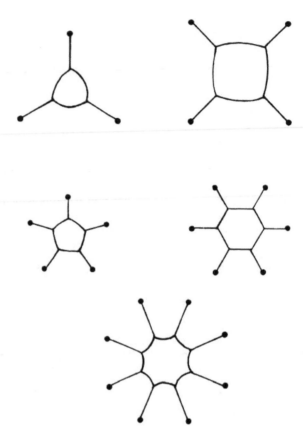

Figure 2.11 Shapes of bubbles arranged within two-dimensional pin patterns.

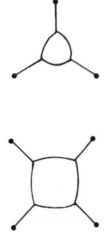

Considering the example with **three pins**, we see that the sides of the bubble curve outwards very distinctly. This is necessary to produce the three internal angles of 120°. The only way that the sides can curve out is if there is an excess air-pressure within the bubble, analogous to the air pressure within a balloon that gives the balloon its shape. The pressure within the bubble needs to be of the requisite value to produce the necessary curvature for the sides and hence the correct angles at the corners. When the framework is first inserted within the soap solution, a certain quantity of air will be trapped. The amount of trapped air, together with the requisite excess pressure required within the bubble to produce the correct curvature, determines the size of the resulting bubble. So if we trap more air, we see a larger bubble, although the corner angles remain 120°.

Moving on to the **four-pin framework**, the required

curvature of the single trapped bubble is less. Curvature is further reduced for the bubble trapped by **five pins**. By the time we have increased the number of pins to six, the situation has changed. A **hexagonal bubble** meets the condition for precise 120° angles, so that a straight-sided hexagonal bubble results. The volume of this bubble will equal precisely the volume of air originally trapped at atmospheric pressure. We can insert the end of a wetted straw into the interior of the bubble and either blow or suck. If extra air is blown in, the bubble increases in size and will ultimately link the six pins. If air is sucked out, the bubble gets smaller. Alternatively, air can be added, extracted, and so on, to see the bubble increase, decrease, and so on in size. The sides remain linear. The bubble changes size with surprising stability, being always centred with respect to the six pins. The six arms are of equal length. (This assumes the framework is held horizontally so that there is no distorting effect from gravity.) It can be shown very easily that as we increase the size of the bubble, the total increase in length of the bubble's perimeter is exactly matched by the total decrease in length of the arms. Theoretically, no energy is added to the film, or extracted from it, in order to alter bubble size. Note that we did not include the pattern involving a hexagonal bubble within the six-pin framework in figure 2.6, as we needed to complete only five sides to join all the pins.

For frameworks involving more than six symmetrically placed pins, the trapped bubbles have sides which curve inwards. Again it is clear that this curvature is necessary to produce the 120° internal angles. As an example, an **eight-pin arrangement** shows the convex sides to the bubble very clearly. The pressure within the bubble is less than air pressure. If we were to drill a small opening through one of the perspex plates in the region of the bubble, air would flow in to equilibrate the pressure and the bubble would expand outwards. In contrast, if we did the same with the bubble trapped by five pins or less, air would flow from the bubble, which would collapse. In the case of the hexagonal bubble, drilling a hole through the perspex to link with the interior of the trapped bubble would alter nothing.

2.4 SYMMETRY OF PATTERNS, OBJECTS AND SOAP FILMS

Many patterns show mirror symmetry. This means that if we position a mirror along an appropriate line within such a pattern, and look into the mirror, we see in that mirror exactly the same

original   ≡   mirror image

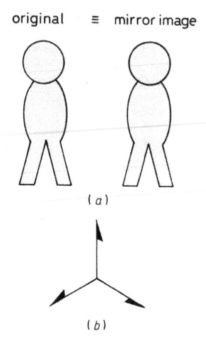

(a)

(b)

Figure 2.12. (a) Mirror symmetry. (b) Rotational symmetry (3-fold symmetry).

pattern as existed in the original (figure 2.12(a)). Other patterns show rotational symmetry. Thus, if we rotate the pattern by some fraction of a complete revolution, we can identify no difference between the pattern in its new configuration and the pattern in its original configuration. Figure 2.12(b) shows a pattern which has three-fold rotational symmetry. It will look the same if rotated by one-third of a complete revolution, i.e. by $2\pi/3$. It will also look the same if rotated by two-thirds of a revolution, i.e. by $4\pi/3$.

Patterns which can be put together to produce a wall-paper design must have rotational symmetry of $2\pi/6$, $2\pi/4$, $2\pi/3$ or $2\pi/2$. Isolated patterns can have rotational symmetry corresponding to 2π divided by any integer we might choose to select. The bubble patterns established within five-pin and eight-pin frameworks (figure 2.11) give examples of such patterns, which could not be reproduced in a repetitive way on wall-paper. However, the hexagonal bubble pattern established within six pins could be used as the basis of a repeating wall-paper pattern. Sometimes analysis of two-dimensional patterns identifies both rotational and mirror symmetry. A very non-symmetrical pattern is one which does not repeat itself unless it is rotated by a full revolution back to its original position. Such a pattern is the soap-film pattern of figure 2.9(c). On the other hand, the soap-film pattern shown in figure 2.4(b) repeats itself every 120°, just as did the example in figure 2.12(b). However, the film of figure 2.4(b) shows higher symmetry than the example in figure 2.12(b). As well as the three-fold rotational symmetry, we see that it has three mirror lines; hence crystallographers describe the pattern as $3m$ to denote the three-fold rotation and the mirrors.

Describing the symmetry of three-dimensional objects and patterns is very important to crystallographers and is a subject in itself. For instance, a cube has four-fold rotational symmetry about any axis joining the centres of opposite faces. It also exhibits three-fold rotational symmetry about any diagonal axis. In addition, the cube has numerous mirror planes. A standard brick as used for house-building has two-fold rotational symmetry, instead of the four-fold of the cube, as none of its faces is square. It also exhibits certain of the mirror planes, but will no longer possess three-fold rotational symmetry.

Crystals can be considered as built from building blocks. For one particular type of crystal the building blocks will be all the same shape and size, but for different types of crystals the blocks will be different. The blocks will possess different symmetry elements. A completely polycrystalline material will have no special axes (it will be analogous to a sphere). It will be isotropic; that is, it will have the same physical properties in any direction. So a polycrystalline sample will be an example of a high-symmetry material. It will be important to appreciate this when considering phase transitions. If a material transforms into a disordered state, it converts into the high-symmetry phase. Similarly, when a soap film changes its pattern it will change its symmetry.

2.5 FURTHER DISCUSSIONS ON GEOMETRY

2.5.1 Intersection of three circles

Given any three circles, each of which intersects the other two, it is a generally known result that their three chords of intersection meet at a point (see figure 2.13(a)). There is no special requirement for these overlapping circles. They do not need to be of equal radius or centred at any special points. However, if we move them until the region overlapped by all three circles diminishes to a point (figure 2.13(b)), then we obtain the vertex for a film joining three pins placed at the other ends of the three chords.

2.5.2 Further interesting equalities

Let us start with any acute-angled triangle ABC. Next, we construct equilateral triangles A'BC, B'CA and C'AB externally on its three sides (see figure 2.14) such that a three-pointed star A'B'C' is formed. Lines AA', BB' and CC' intersect BC, CA and AB, respectively, and in each case the intersection occurs internally.

We now obtain the circumcentres D, E and F of the triangles A'BC, B'CA and C'AB, respectively. As we have seen already, the circumcentre of a triangle is the centre of a circle which passes through its three corners. To obtain such a centre, we perpendicularly bisect the three sides of the triangle, and find the point of intersection of these perpendiculars.

This construction produces some remarkable results:

(a) Points D, E and F form the vertices of an equilateral triangle.

(b) $AA' = BB' = CC' = \sqrt{3}EF = \sqrt{3}FD = \sqrt{3}DE$.

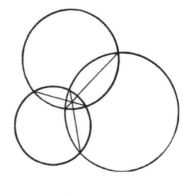

(a)

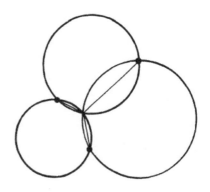

(b)

Figure 2.13. Three overlapping circles. (a) General case. (b) Meeting at a common point.

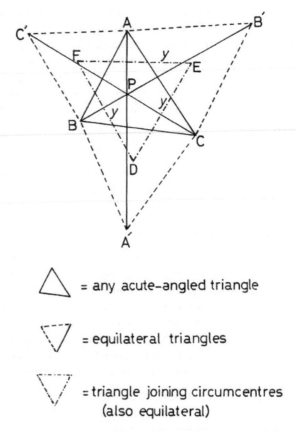

Figure 2.14 Illustration of some remarkable geometrical results involving triangles.

(c) AA′, BB′ and CC′ meet at point P within the interior of triangle ABC. P is called the Fermat point.

(d) The angles $B\widehat{P}C$, $C\widehat{P}A$ and $A\widehat{P}B$ are all equal to 120°. (Hence one sees the connection with soap films.) These angles are bisected by AA′, BB′ and CC′, respectively.

(e) AP + BP + CP = AA′ = $\sqrt{3}$EF and so on. If we let the sides of the equilateral triangle DEF be y, then AP + BP + CP = $\sqrt{3}y$. This is the same relationship which was shown in mathematical box 2.1 on pages 18–19. AP + BP + CP forms the minimum network linking A, B and C (see box 2.2, page 28).

(f) EF perpendicularly bisects AP; FD perpendicularly bisects BP; and DE perpendicularly bisects CP.

2.5.3 The three-point Steiner problem

The three-pin soap-film pattern demonstrates empirically how to join three points by a path of minimum length. Jacob Steiner (1796–1863), a famous geometrician, looked at the problem mathematically. From the Steiner problem (see box 2.2), we see why the 120° angles always occur between film components.

2.6 THE GENERAL STEINER PROBLEM

The 120° angle will be repeated many times in complicated soap film patterns. We can see from the earlier figures (for example figure 2.2) how a generalized Steiner problem of joining four or more points can lead to two or more possible solutions. There is no general proof for solving the problem of path configuration and overall minimum path length to join n points but we do see that the paths must meet at vertices at 120°. If we have n starting points, then the maximum number of intersections required to link these points is $(n-2)$, but the number of intersections may take any value in the range 0 to $(n-2)$ depending on the original geometrical arrangement of the starting points.

The tree of minimum length to join n points is called a Steiner minimum tree. Steiner obtained such networks in simple cases by introducing new points which he then used as vertices, and so on. As we already know, the Steiner minimum tree for three pins arranged at the corners of an equilateral triangle involves one extra vertex at the circumcentre of the triangle.

Computer scientists tackled the problem by finding the so-called minimum spanning tree which has no extra vertices. Thus for the case of the three pairs forming an equilateral triangle, the minimum-spanning-tree solution is a V connecting one pin directly to the other two. Similarly in this method, one goes on to join up other pins in a more complex arrangement and the minimum-spanning-tree solution will be that arrangement which joins all the pins according to this convention, with no extra junctions, and is the minimum length as obtained by this convention. It is comparatively simple to check all the possibilities using a computer (although even this problem becomes time consuming as the value of n becomes large), whereas it is not possible to check possibilities if we are required to insert an arbitrary number of extra vertices (but less than $n-2$) in arbitrary positions, as we would need to do for the Steiner solution. Unless the points being joined are on some form of equilateral grid, the minimum-spanning-tree solution will be

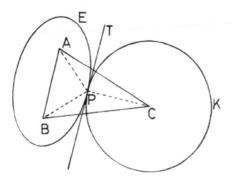

Figure 2.15. Illustrating the Steiner problem to find the shortest path to join three points.

Box 2.2   Solution of the three-point Steiner problem

Let us take three points A, B and C, arranged in any way in a plane. The problem is to find a fourth point P such that the sum of the distances from P to the other three points is a minimum; that is we require $AP + BP + CP$ to be a minimum length. We can perhaps guess the solution from what we have seen already. If the triangle ABC possesses internal angles which are all less than 120°, then P is the point such that each side of the triangle, i.e. AB, BC and CA, subtends an angle of 120° at P. However, if one angle, say $A\widehat{C}B$, is greater than 120°, then P must coincide with C. The latter alternative is obvious. P cannot sit inside such a triangle with the three sides subtending 120°, and it must lie at C rather than at the other points because $CA + CB$ is less than the sum of the lengths of any other two sides of the triangle ABC. So we are left to prove the former alternative.

We draw an ellipse E through the point P with foci A and B (figure 2.15). Any point on the ellipse, including point P, has the property

$$AP + BP = l$$

where l is a constant. This is a well-known property of an ellipse. It is used as the basis of a common method of drawing an ellipse. The two ends of a fixed length l of string are pinned down at the foci A and B. A pencil is inserted loosely against one side of the string and the ellipse drawn by keeping the string taut around the lower end P of the pencil. $AP + BP$ remains constant, and an ellipse results.

Next we draw a circle K, radius CP, and centred at C. If point P is to meet our condition of $AP + BP + CP$ being a minimum, then we can draw through P a tangent T common to both the ellipse E and the circle K. T must be tangent to the ellipse E if $AP + BP = l$ is a minimum. Otherwise, a larger value of $AB + BP$ would produce an ellipse which would intersect the circle twice. It is the property of an ellipse that AP and BP are equally inclined to the tangent T; hence, angles $A\widehat{P}C$ and $B\widehat{P}C$ are equal.

Box 2.2 *Continued*

We can now repeat the arguments by taking the foci of the ellipse as B, C and C, A and the centres of the circle as respectively A and B.

From this we can see that P must be positioned so that $AP + BP + CP$ is a minimum and $A\widehat{P}C = B\widehat{P}C = A\widehat{P}B$.

Also, since

$$A\widehat{P}C + B\widehat{P}C + A\widehat{P}B = 360°$$

all three angles must be 120°.

---

longer than the Steiner minimum tree. The ratio of the minimum-spanning-tree length to the Steiner-minimum-tree length is called the Steiner ratio and for the three-pin case cited the ratio is $\sqrt{3}/2$ (or approximately 0.866).

Gilbert and Pollack (1968) conjectured that the ratio never falls below $\sqrt{3}/2$ so that the Steiner minimum tree can never be better than the minimum spanning tree by more than approximately 13%. What the precise ratio should be is subject to some discussion but it cannot fall below 0.824 for up to seven pins. The problem has now been looked at for n points by Du and Hwang (1990), and they have shown that the conjecture is correct for n points. They used game theory techniques and a minimax approach as used in operational research.

3

Soap films and first- and second-order phase transitions

In this chapter are described two-dimensional models in which a pair of pins move, or alternatively a single pin moves, whilst other pins remain fixed. These models demonstrate changes of film pattern that are analogous to first-order transitions, as demonstrated by liquid–vapour and solid–liquid changes, and second-order phase transitions, as typified by the order–disorder phase transition in β-brass. The thermodynamics of these transitions is discussed.

3.1 A 2D MODEL FOR A FIRST-ORDER PHASE TRANSITION

In the previous chapter, we discussed two-dimensional soap-film patterns and showed that it is relatively easy to calculate total film lengths. Thus, two-dimensional models are useful for finding the actual variation of soap-film energy as we alter the relative positions of the pins. We can identify and calculate the energy changes for transformations in the film patterns. In addition, we shall see how these changes link with the thermodynamics of phase transformations.

We consider a model consisting of four pins, A, B, C and D, arranged at the corners of a rectangle. The lengths of the sides AC and BD are variable (see figure 3.1). If we start with the situation in which sides AB (=CD) and AC (=BD) are equal, such that the pins sit at the corners of a square, then the soap film can take up two configurations, as was discussed on page 13. These two configurations are shown in the figure. One configuration is shown with continuous lines and the other is shown with dashed lines.

Suppose we keep pins A and B fixed, and move pins C and D outwards and inwards together, as shown by the arrows. It

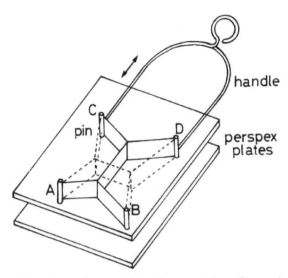

Figure 3.1 Four-pin model to demonstrate a first-order phase transition.

is easy to see that the film will switch backwards and forwards between the two configurations. Provided movement of pins C and D is slow, the film alters such that it always maintains its equilibrium shape, a shape which corresponds to minimum length. The complete range of changes is shown in figures 3.2(a) to 3.2(g).

3.1.1 Stages in altering the shape of the first-order soap-film model

3.1.1.1 Commencement with small pin separation. The first figure (3.2(a)) shows a pair of pins AB close to a pair of pins CD, giving rise to the dashed line configuration in figure 3.1. At equilibrium, each component of the soap film meets two other components at a vertex at 120°. Hence, calculation of the total film length for $AB = CD = a$ (fixed) and $AC = BD = x$ (variable) is straight-forward. If we were to blow the film, it would distort and its energy increase, before relaxing back to its equilibrium shape and position. This is shown in the energy (film length) versus configuration curve where the dot represents the equilibrium energy or equilibrium position of the film. Blowing the central region of the film towards AC (say) takes the film up one side of the energy curve, and blowing it towards BD (say) takes it up the opposite side of the energy well. The shape of the well will be different in different directions.

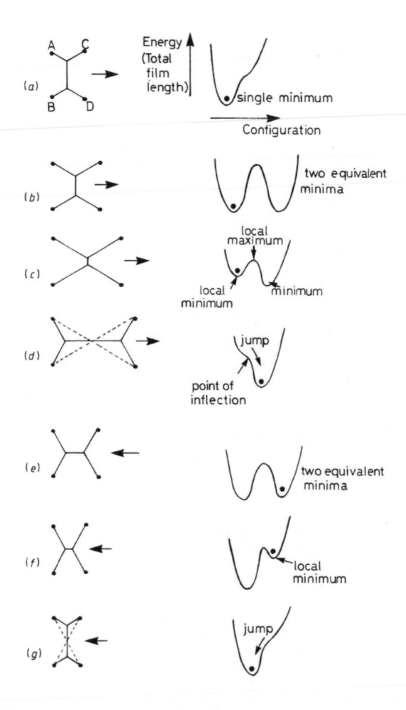

Figure 3.2 Film length (energy) versus configuration plots for films patterns for the four-pin model shown in figure 3.1. For a description of the stages (a) to (g) see text.

This is shown diagrammatically in the energy–configuration curve, without the diagram telling us what specific movement is represented by a particular change in the configuration parameter. Later we will suggest what variable we might use for this configuration parameter, but we should note that it is measuring a deviation from equilibrium.

3.1.1.2 Increasing the pin separation.   Now we slowly move pins C and D away from A and B such that x increases. There may be a tendency for some sticking of the film, but the film will gradually take up its equilibrium shape. The corresponding energy versus configuration curve will change shape as pins C and D are moved outwards, but the film retains an equilibrium shape consistent with sitting at the bottom of the well. At the stage when the four pins occupy the four corners of a square, the film will still be occupying the bottom of the energy well. However, another well of equal depth will have become established alongside (figure 3.2(b)). This second well corresponds to the alternative soap-film configuration. The film cannot jump into this other well. It would require a substantial amount of energy in order to surmount the energy hill between. We move C and D further outwards. Still the film sits in its original energy well. This is despite the fact that the alternative configuration now has a lower energy. The film is sitting in a local minimum. The situation is shown in figure 3.2(c), where the film cannot jump into the lower energy well because of the small hillock (local maximum) remaining between the wells. Of course, it would not now require much perturbation of the soap film to induce it to switch into the lower minimum. Pulling C and D even further outwards allows us to reach a separation such that the two vertices within the original soap film come into contact. The film instantly switches to its other configuration and into the other energy well. This happens when the original energy well ceases to be identifiable, the hillock having changed into a point of inflection on the energy versus configuration graph. It is as if the soap film can now spill over into the other minimum (figure 3.2(d)). We can continue pulling the pins C and D outwards, but nothing further is achieved other than to increase the overall film length, and to make the single energy minimum more pronounced.

3.1.1.3 Reversing the motion and reducing pin separation.   Let us reverse the direction of movement of C and D. The first notable position is when the pins have returned to the corners

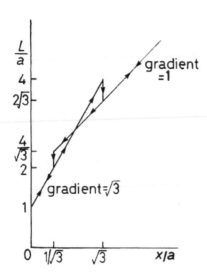

Figure 3.3. Variation of film length L (or energy) with control variable x for the first-order phase-transition model.

of a square. We have the identical energy-configuration curve to that shown in figure 3.2(b). However, this time the soap film is sitting in the alternative energy well (see figure 3.2(e)) to that in which it started. Further reduction of the pin separation raises the minimum in which the film is sitting, but once again the film cannot jump the barrier to the lower minimum (figure 3.2(f)). Eventually, we return to a position where the two vertices of the film touch. A new point of inflection in the energy versus configuration curve is obtained. The film switches to the original configuration and re-enters the original energy well (figure 3.2(g)).

3.1.2 Calculation of film length

It is important for much that follows that we calculate the variation of film length as we vary the separation x between AB and CD. We can regard x as the **control variable** (or **control parameter**). By summing the five component lengths together, we can calculate the total length L of the film for any separation x. Figure 3.3 shows this variation of length with separation x once the complete range of calculations has been carried out. We now look at the various steps.

3.1.2.1 Small pin separation. We start off with the pins A,B and C,D very close together such that x is small (figure 3.4(a)). There are four arms each of length $x/\sqrt{3}$ and a central bridge of $(a - x/\sqrt{3})$. This gives a total length L for the film given by

$$L = (a + \sqrt{3}x).$$

The film remains in this configuration until the central bridge tends to zero length, and the two vertices touch. This occurs at $x = \sqrt{3}a$, when the total length of the film, L, is $4a$.

3.1.2.2 Film length after the transition. At this point, the film jumps into a configuration with four arms of length $a/\sqrt{3}$. The central bridge is of length $x - a/\sqrt{3}$ which equals $2a/\sqrt{3}$ at this position of jump. The total film length L is now given by

$$L = (x + \sqrt{3}a).$$

The total film length has suddenly decreased to $2\sqrt{3}a$ or $3.42a$ and the film looks like that in figure 3.4(b). We can use the algebraic expression for L to calculate the film length for increasing values of x as we move CD further outwards.

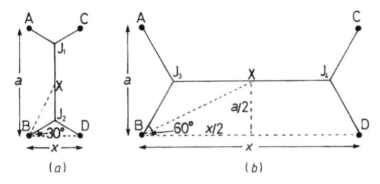

Figure 3.4 Equilibrium soap-film configurations within the four-pin model. (a) Pairs of pins close together. (b) Pairs of pins widely separated.

3.1.2.3 Film length as separation is now reduced. We stop moving pins C and D outwards, and move the pair of pins back towards AB. On decreasing x, it is necessary to move C and D back beyond the point where the transition, just described, first occurred. C and D need to be moved until the two vertices, as shown for this configuration in figure 3.4(b), have moved together and touch. This only occurs when the separation x between AB and CD equals $a/\sqrt{3}$. This position corresponds to a total film length of $4a/\sqrt{3}$ or $2.31a$, and the length reduces at the jump to $2a$.

3.1.2.4 Combining the data. Putting the data together provides the complete graph already shown in figure 3.3. Film length L is a direct measure of the energy of the film. There is a portion of the graph which has a gradient, dL/dx, equal to $\sqrt{3}$. This corresponds to the part of the graph where the total film length L is given by $(a + \sqrt{3}x)$. There is another portion of the graph with a gradient of 1, where L is $(x + \sqrt{3}a)$. The variation of dL/dx is shown in figure 3.5.

The change in gradient from $\sqrt{3}$ to 1 occurs as the soap-film pattern changes configuration. The path taken as shown by both figures 3.3 and 3.5 is different according to whether x is increasing or decreasing. This is the phenomenon of hysteresis. The path difference in the two directions is a direct consequence of the requirement that the two vertices of the film meet for pattern switching to occur. As we have seen, this occurs for two different values of x, one during expansion and the other during contraction. As suggested earlier, the two portions of

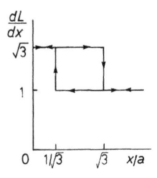

Figure 3.5. Variation dL/dx with x for the first-order phase-transition model.

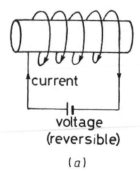

current

voltage
(reversible)

(a)

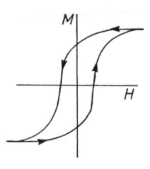

(b)

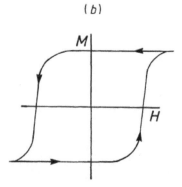

(c)

Figure 3.6. Hysteresis in magnetic materials. (a) Set-up for producing magnetization. (b) and (c) *M–H* loops in soft and hard magnetic materials respectively.

Box 3.1   Hysteresis loops in magnetic materials

Such so-called hysteresis loops are also observed when magnetizing certain metals and metal alloys, such as iron and steel. The metal is inserted in the centre of a current-carrying coil (figure 3.6(a)). The magnetization M is plotted as a function of the magnetic field H produced by the current through the coil. The resulting curves can take the extreme forms shown in figures 3.6(b) and (c), or alternatively some intermediate form. Figure 3.6(b) applies to a soft material which is easily magnetized and figure 3.6(c) is applicable to a hard magnetic material which is not so easily magnetized. We notice the considerable similarity between the hysteresis loop for magnetization in a hard magnetic material and that for the hysteresis loop for a soap film. The area within the MH loop for the magnetic material measures the energy dissipation when the magnetic field H is cycled once, just as the area within the soap-film hysteresis rectangle represents the loss of energy for a single expansion and contraction of the soap-film model. When magnetizing a hard material, nothing happens until a high magnetizing field is applied. The atoms, which act rather like little magnets, all become aligned and the magnetization M suddenly increases. On decreasing the magnetizing current and even reversing it, no change occurs to M until the field H is of sufficient magnitude to reverse the alignment of the atoms. (In reality, it is the rotation of the outer unpaired electrons of the atoms of the metal which produces the magnetization, and it is this rotation which must be reversed.)

the graphs can be considered as equivalent to the presence of different phases of a crystalline material on either side of a transition temperature. The area within the hysteresis rectangle in figure 3.5 is equivalent to heating of the film produced by the nett work done in causing the film to expand and contract and so measures the energy dissipated. What we are seeing is the modelling of a first-order phase change, with distinct jumps in both energy and volume at the transition. For phase changes in materials, the jump in energy is referred to as latent heat.

3.1.3 Introducing the configuration parameter

Figure 3.2 showed curves for the variation of energy versus configuration for different separations of the pairs of pins A,B and C,D as we blew the film *away from its equilibrium configurations*. We can calculate such curves by choosing a suitable displacement x of CD relative to AB and an appropriate configuration parameter. We assume that blowing displaces the junction points J_1 and J_2 in figure 3.4(a), and the junction points J_3 and J_4 in figure 3.4(b), by equal amounts either towards or away from the centre X of the figures. If the junctions are blown towards each other, they will eventually touch and then move apart in the alternative configuration. There will be a smooth change of overall length as this happens.

The first of the two configurations shown in figure 3.4 is redrawn in figure 3.7(a) to (c) but with the angles at the junction points J_1 and J_2 no longer equal to 120°. The junction points have been moved symmetrically either towards or away from the centre point X, including moving beyond X. An angle β is defined with its apex at one of the four pins, say pin B. The angle is that lying between BX and BJ_2 (or BJ_3). We note that BX lies along a diagonal joining diagonally opposite pins, and is fixed. BJ_2 or BJ_3 joins the chosen pin to the junction point, and varies its direction. However, we choose the square root of β for the *configuration parameter*.

Looking again at figure 3.7, (a) to (c), we see that the β angles are indicated in anticlockwise and clockwise directions. So one direction must be negative. Nevertheless, we must take

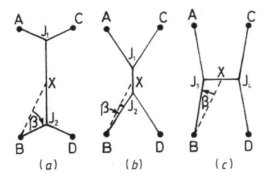

Figure 3.7 Soap-film configurations within the four-pin model away from equilibrium for the case of pairs of pins close together. Figures (a) to (c) show configurations with different values of the configuration parameter β.

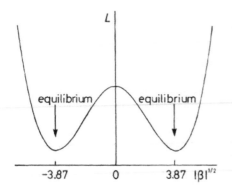

Figure 3.8. Variation of film length with configuration parameter $|\beta|^{1/2}$ as we blow the film away from equilibrium. (Pins A, B, C and D are assumed to be at the corners of a square.)

the square root of a positive angle so we take β as positive in either direction. However, we attach a negative sign to $\beta^{1/2}$ when β is measured anticlockwise, and a positive sign when β is measured clockwise.

If we take a particular separation x for the pairs of pins, i.e. we take a fixed value for the control variable, we may plot film length L as it varies with β. An appropriate example is to take $x = a$ (i.e. the pins lie at the corners of a square); BX is at 45° to BD. At equilibrium, BJ_2 is *either* at 30° or at 60° to BD. This means that the minimum film length (and energy) occurs at positions of $-[15]^{1/2}$ and $+[15]^{1/2}$, i.e. ± 3.87, if we measure angles in degrees (figure 3.8). As expected, the minima are symmetrically displaced from the central maximum, which occurs when the film junctions touch at the centre of the square. This central maximum represents unstable equilibrium. With x unequal to a, the two minima are not symmetrically positioned with respect to $\beta = 0$. A range of plots of L versus $\pm|\beta|^{1/2}$ was shown in figure 3.2. *Mathematica* programs A.II.1 and A.II.1a in Appendix II demonstrate these curves. Program A.II.1a includes animation.

3.2 THERMODYNAMICS OF THE PHASE TRANSITION

The changes of pattern and energy of the soap film can be described in terms of thermodynamics and can be related to many other physical changes which are described by thermodynamics. The *First Law of Thermodynamics* is the appropriate starting point. It states that a change in the internal energy of a system is equal to the amount of heat absorbed less the amount of work performed by the system. The small amount of heat input to the system, at absolute temperature T, can be expressed in terms of the corresponding change of entropy; i.e., it can be expressed as $T\,dS$ where dS is the incremental change of entropy. Thus, if we have an amount of heat ΔQ going into our system at temperature T the change of entropy is $\Delta Q/T$. An alternative method of describing entropy is to relate it to the amount of disorder in a system. The larger the entropy of a system, the more disordered or muddled is that system. As the age of the universe increases, its entropy inexorably gets larger as overall disorder increases also. In describing soap films, we are considering the patterning, so it is not surprising that entropy and order are involved.

If we call the incremental change of internal energy dU and we say an incremental amount of work Δ(work) is performed by the system, the First Law can be given in the incremental (differential) form

$$dU = T\,dS - \Delta(\text{work}).$$

In thermodynamics, we usually first consider a fluid system at pressure P. If the system does work by pushing a piston against pressure P to produce a change of volume (dV), then:

$$\Delta(\text{work}) = P\,dV$$

and the first law becomes

$$dU = T\,dS - P\,dV.$$

Whereas a fluid system exists in three dimensions, a soap film is two-dimensional. It is necessary to replace the pressure P for a fluid with surface tension $-\gamma$ for the soap film, or -2γ for a two-sided film, and to replace the change of volume dV with change of surface area dA. Remember that γ is a tension, and this is the reason for the negative sign as shown. Thus, the First Law of Thermodynamics for a two-sided film becomes

$$dU = T\,dS + 2\gamma\,dA.$$

This equation for the First Law involving dU is fundamental, but it turns out that internal energy U is itself not very appropriate for explaining changes in systems. Thermodynamicists have developed the use of quantities called free energies. There are two important ones that apply to fluid systems and these are called the **Gibbs free energy** and the **Helmholtz free energy**. The Gibbs free energy G is defined as

$$G = U - TS + PV.$$

An increment of Gibbs free energy dG will be related to increments of the other quantities by the equation

$$dG = dU - S\,dT - T\,dS + P\,dV + V\,dP.$$

We now substitute the expression for dU as given by the First Law and obtain

$$dG = -S\,dT - V\,dP$$

which shows that we have G as a function of T and P. In the laboratory we usually alter the temperature or pressure within our system.

The incremental form of the Helmholtz free energy F is similar to that for the Gibbs free energy and is given by

$$dF = -S\,dT - P\,dV.$$

Change of volume V in a fluid system is analogous to change of length x, and hence change of area, in the two-dimensional soap-film system. Looking at the equations, we can immediately see a symmetry (and hence an analogy) between change of temperature T, pressure P, and volume V within the equations for change of free energy.

Let us return to the equation for the differential form of the Gibbs free energy dG and suppose that we keep the pressure constant. This means $dP = 0$ and we have

$$dG = -S\,dT.$$

Dividing both sides by dT gives us an expression for entropy:

$$S = -\left(\frac{\partial G}{\partial T}\right)_P$$

The subscript P is added to indicate that pressure is held constant. We have used ∂ to represent a small change as applied to one parameter when another parameter (in this case P) has been kept constant.

Now consider an equation for a small change in entropy S of our system. If S increases slightly, this means a slight increase in disorder. To obtain the change we require the second differential and our equation becomes

$$dS = -\left(\frac{\partial^2 G}{\partial T^2}\right)_P dT.$$

As $\Delta S = \Delta Q/T$ and the specific heat at constant pressure C_P is $\Delta Q/\Delta T$, then

$$\frac{\Delta S}{\Delta T}\left(=\frac{dS}{dT}\right) = \frac{C_P}{T}.$$

Hence we have

$$C_P = -T\left(\frac{\partial^2 G}{\partial T^2}\right)_P.$$

The nett result of the above is that we have the Gibbs free energy defining the energy of the system, the first differential of the Gibbs free energy defining the entropy of the system, and the second differential defining the specific heat. Just as the soap film attempts to minimize its energy, a thermodynamic system tends towards that state where its free energy is minimum.

3.3 THE WATER–VAPOUR PHASE TRANSITION

In order to examine phase transitions, we should consider a one-component system, say H_2O molecules, which can exist in two phases in contact, say the water and water-vapour phases. At certain temperatures and pressures, all the H_2O molecules will be in the water phase, and at other temperatures they will be in the vapour phase. If we keep the pressure constant at atmospheric, then all the H_2O molecules will be in the form of water at temperatures below 100 °C and will be vapour at temperatures above 100 °C. There will be an intermediate temperature around 100 °C where the water and vapour can coexist in equilibrium. At this equilibrium between the two phases, the specific Gibbs function (Gibbs free energy per unit mass) must be the same. Otherwise there will be transformation of mass from the phase with higher specific free energy into the phase with lower specific free energy.

If we heat water above 100 °C, the liquid should vaporize; this is because the vapour will now have the lower specific free energy. What we may see however is a delay in the vaporization. This delay arises when extra energy is required to start the vaporization process. The phenomenon is called superheating. There is a need to provide sufficient energy to produce a bubble of vapour of finite size, often at the boundary of the container. The effect can be even more distinctive in the reverse direction, when supersaturation may occur before vapour molecules can produce liquid drops. Hence, a phase change commonly happens at a temperature which is higher for vaporization than that calculated for equal specific energies between the two phases, and at a temperature which is lower for liquefaction.

An alternative approach to altering temperature is that of altering pressure. For example, we know that water boils at a lower temperature if we lower the pressure. It is common knowledge that as water boils at a lower temperature on a high mountain, it becomes less easy to produce a good cup of tea!

We can illustrate the phase change produced by changing the temperature using graphs in which the variations of G, S and C_P are plotted as functions of temperature T. These graphs are shown in diagrammatic form in figure 3.9. Here, figure 3.9(a) shows the variation of Gibbs free energy G with temperature and it should be noted that the free energy *decreases* with increasing temperature. This figure is similar to figure 3.3 for the variation of L with x for the soap film except that *decreasing* temperature

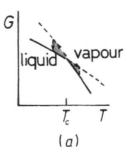

(a)

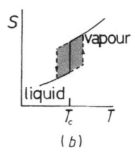

(b)

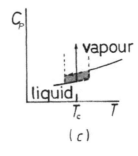

(c)

Figure 3.9. First-order phase transition. Variation of G, S and C_P with temperature T.

is equivalent to *increasing* x. The dashed lines and the hatched areas correspond to superheating and supercooling. Figure 3.9(b) shows the change of entropy in the two phases with the much greater entropy in the vapour phase. The jump in entropy at the phase transition equals

$$\text{heat input} \times \text{temperature}$$

and hence is

$$\text{latent heat} \times \text{temperature}.$$

The hatched area corresponds to the heating (hysteresis effect) due to superheating and supercooling. Figure 3.9(c) shows the variation in the specific heat including the fact that it goes infinite for a sudden phase transition when heat input is producing reordering but no temperature change.

3.4 PHYSICS OF THE FOUR-PIN SOAP-FILM MODEL

Let us choose the Helmholtz free energy, F, and take the differential form $dF = -S\,dT - P\,dV$. If we assume temperature is constant,

$$P = -\left(\frac{\partial F}{\partial V}\right)_T.$$

For a fluid the first differential is with respect to volume, whereas for the film it is with respect to the parameter x. In place of pressure we involve surface tension, but it is a resolved component equal to the force on the moving pins (see box 3.2). The second differential gives the elastic modulus. dL/dx is constant for both configurations, or phases, for the four-pin model. The elastic modulus, d^2L/dx^2, is also zero for both phases as no change of force is required to produce an increase in area in either phase.

In Appendix II, *Mathematica* program A.II.2 animates the movement of the pins and illustrates the corresponding variation of L with x. Program A.II.3 shows the pin animation but this time illustrates the corresponding variation of dL/dx.

Assuming the surface tension in the film is γ, that the film has two surfaces, and that the height of each pin is h, the force in the direction x required to keep pins A,B and C,D a distance x apart is $2\gamma h(dL/dx)$ as is shown in box 3.2.

Thus on the low-x side, a movement of the pins by a distance x produces a change of energy equal to the force times the distance, that is $2\sqrt{3}\gamma hx$. This implies that the area of the film

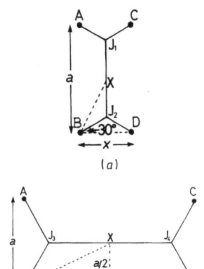

Figure 3.4. *Revisited.* Equilibrium soap-film configurations within the four-pin model. (a) Pairs of pins close together. (b) Pairs of pins widely separated.

Box 3.2 Force between pins

We can calculate the force required to keep the pins apart in the x direction and hence the force to move them apart infinitesimally slowly. In the equilibrium configuration with x small (figure 3.4(a)), the force in the x direction on the pins C,D is

$$\mathcal{F}_x = 2\gamma h \cos 30° \times 2 = 2\gamma h \sqrt{3}.$$

For this film pattern, $L = a + \sqrt{3}x$, so that

$$\frac{\mathrm{d}L}{\mathrm{d}x} = \sqrt{3}.$$

Hence,

$$\mathcal{F}_x = 2\gamma h (\mathrm{d}L/\mathrm{d}x).$$

We would expect this from the fact that the change of energy will be

force × distance

equal to $\mathcal{F}_x\,\mathrm{d}x = 2\gamma\,\mathrm{d}L$.

For the equilibrium configuration with x large (figure 3.4(b)), the force in the x direction is

$$\mathcal{F}_x = 2\gamma h \cos 60° \times 2 = 2\gamma h$$
$$= \text{force between } J_3 \text{ and } J_4.$$

Here, $L = x + \sqrt{3}a$ so that $\mathrm{d}L/\mathrm{d}x = 1$, and again

$$\mathcal{F}_x = 2\gamma h (\mathrm{d}L/\mathrm{d}x).$$

---

has changed by $\sqrt{3}hx$. A quick check on the geometry shows that the four arms of the film have altered their total area by this amount (each single arm changes its area by $(\sqrt{3}/4)hx$). The situation for the high-x phase is even more obvious. The change of energy is $2\gamma hx$. The total area of the arms is constant; the area of the central component only is altered by hx.

If one follows the rectangular hysteresis loop in figure 3.5, the work done is the area within the hysteresis loop, i.e.

$$2\gamma \times (\mathrm{d}L/\mathrm{d}x) \times h \times \mathrm{d}x.$$

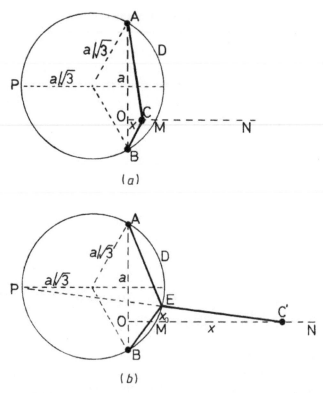

Figure 3.10 Three-pin model to demonstrate a second-order phase transition.

An appropriate value for the surface tension of the soap (or detergent film) is 25 mN m^{-1}. The value of $d(dL/dx)$ is $\sqrt{3}-1$ or 0.73, and, for the size of model envisaged, h will be 10 mm and dx say 6 cm. This gives rise to a change of energy of 22 mJ. Assuming the specific heat of the film to be approximately that of water (4.2 kJ kg^{-1} K^{-1}) and assuming a film thickness of say 0.3 μm (see page 2), one cycle of the film under adiabatic conditions should produce a change of temperature of the order of 10^{-2} K. (See box 3.3.)

3.5 A 2D MODEL FOR A SECOND-ORDER PHASE TRANSITION

A very different kind of transition occurs if we consider only three pins as shown in figures 3.10(a) and (b). Pins A and B are fixed and pin C has variable position along a line perpendicular

---

Box 3.3   Change of soap-film temperature due to work

Work done $= 2\gamma \times d(dL/dx) \times h \times dx$
$$= 2 \times 25 \times 10^{-3} \times (\sqrt{3} - 1)$$
$$\times 10^{-2} \times 6 \times 10^{-2} \text{ J}$$
$$= 2.2 \times 10^{-5} \text{ J}.$$

If t is the film thickness, C_P the specific heat of the film, ρ the density and ΔT the temperature change,

thermal energy $= L \times h \times t \times C_P \times \rho \times \Delta T$
$$= 16 \times 10^{-2} \times 10^{-2} \times 0.3 \times 10^{-6}$$
$$\times 4.2 \times 10^{3} \times 10^{3} \times \Delta T \text{ J}$$
$$= 2.0 \times 10^{-3}\Delta T \text{ J}.$$

If all the work is converted to thermal energy, $\Delta T \sim 10^{-2}$ K.

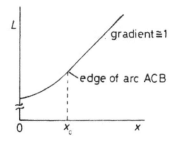

Figure 3.11. Variation of film length L (or energy) with control parameter x for a 3-pin model (second-order model).

to AB, shown as the dashed line OMN. When the movable pin lies inside the arc ADB (figure 3.10(a)), the equilibrium film will consist of only two lengths AC and CB. When this pin lies outside the arc at a position C′, the film will consist of lengths AE, EB and EC′, meeting at E on arc ADB at 120° angles. The distance of the movable pin from O is taken as x. When the pin is outside the arc ADB at C′, the total film length will equal C′P, which arises from the geometrical equivalence shown on page 18.

3.5.1   Variation of film length (energy) with control parameter

Again we calculate and plot the variation of total film length L with x, which is the control parameter. The curve appears smooth as shown in figure 3.11. It is only when we plot the gradient dL/dx as a function of x (figure 3.12) that we obtain a sudden change, the feature which we might regard as characteristic of a phase transition. This change occurs at $x = x_0$, where x_0 measures OM, the value of x on the arc ADB. The fact that we need to go to the first differential before we see this

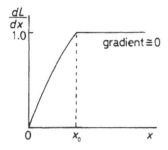

Figure 3.12. Variation of dL/dx with x for a 3-pin model.

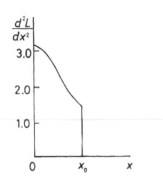

Figure 3.13. Variation of d^2L/dx^2 with x for a 3-pin model.

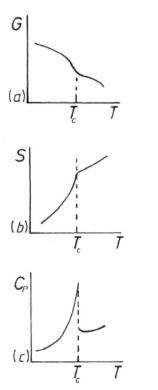

Figure 3.14. Thermodynamic characteristics for a second-order transition. Variation of G, S and C_P with temperature T.

sudden change (compare figure 3.12 with figure 3.3) means that we are observing the analogue of a so-called second-order phase change.

The graph showing the variation of dL/dx is analogous to a plot of the first differential of specific Gibbs free energy G with respect to temperature (but again note decreasing temperature is equivalent to increasing x). Compare the graph with that for the first-order transition (figure 3.5). In physical terms, dL/dx again represents the force on the control pin C as it is moved. In this model, the force is not constant even for the second film configuration because the film angle at the control pin C alters as the pin moves along ON. However, this variation is too small beyond arc ADB to be easily identified in figure 3.12. d^2L/dx^2 (figure 3.13) represents the elastic modulus of the model and this also varies very slowly with x when the pin C is outside arc ADB.

Although it is relatively easy to calculate values of L for the complete range of values of x, and to plot curves for L and dL/dx versus x, the actual values and shapes of the plots depend on the chosen starting position O of the movable pin along AB. For the special case of O at the midpoint of AB, EC' always lies perpendicular to AB and there is no change of force once the pin moves outside arc ADB. This results in a zero gradient for the dL/dx versus x curve in the region beyond x_0. Nevertheless, work needs to be done to move the pin outwards.

Mathematica program A.II.4 in Appendix II shows animation of the pin movement together with the variation of length of soap film L with separation x. The next program, A.II.5, shows the animation together with the variation of dL/dx with x. Program A.II.6 shows animation plus variation of d^2L/dx^2 with x.

3.6   THE EHRENFEST CLASSIFICATION

The shapes of these curves for the 3-pin model are similar to those for typical second-order transitions as shown in figure 3.14. The curves in this figure should be compared with those in figure 3.9 for first-order transitions. There is some variability in the shapes of the curves according to application. The classification for phase transitions was introduced by Ehrenfest (1880–1933). In this classification, the order of a transition is defined as the lowest order of differential of the Gibbs free energy function G with temperature T that shows a discontinuity at the transition (see table 3.1). So in the second-order transition we do not see a discontinuity in entropy S. This means that

Table 3.1 Thermodynamic quantities in first- and second-order phase transitions.

Order of phase transition	Discontinuity appears in differentials	Corresponding experimental quantities
Thermodynamic quantities		
first	$\left(\dfrac{\partial G}{\partial T}\right)_P$	entropy; $S = -\left(\dfrac{\partial G}{\partial T}\right)_P$
second	$\left(\dfrac{\partial S}{\partial T}\right)_P = -\left(\dfrac{\partial^2 G}{\partial T^2}\right)_P$	specific heat at constant pressure; $C_P = T\left(\dfrac{\partial S}{\partial T}\right)_P$
Soap films		
first	$\left(\dfrac{\partial L}{\partial x}\right)_T$	force; $\mathcal{F}_x \propto \left(\dfrac{\partial L}{\partial x}\right)_T$
second	$\left(\dfrac{\partial \mathcal{F}_x}{\partial x}\right)_T \propto \left(\dfrac{\partial^2 L}{\partial x^2}\right)_T$	elastic modulus

second-order phase changes do not exhibit latent heats. (Nor by analogous arguments do they show a change of volume at the transition.)

A good example of a second-order transition is the order–disorder transition which occurs at approximately 460 °C in β-brass, a 50:50 copper–zinc alloy. In the low-temperature phase the copper and the zinc atoms are arranged in a regular way, whereas in the high-temperature form they are distributed randomly over the lattice sites. β-brass is body-centred cubic. So one set of atoms, say the copper, can be considered to sit in the main cubic framework, and the other, in this case the zinc, sits in the body-centred positions, which are equivalent to a second (displaced) cubic framework. β-brass exhibits a finite specific heat at the phase transition (figure 3.15) and no latent heat.

Other examples of second-order transitions are the transition of certain conducting metals to a superconducting state (although this transition must occur in zero magnetic field for it to be second order) and that of ordinary liquid helium to its superfluid state.

A summary of selected thermodynamic quantities and a comparison with the situation for a soap film are shown in table 3.1.

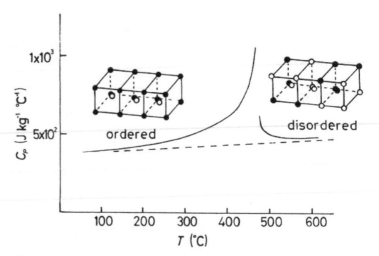

Figure 3.15 The variation of specific heat with temperature for β-brass.

3.7 THE ORDER PARAMETER AND SYMMETRY

A second-order transition involves a lowering of symmetry and it is useful to associate an order parameter with the transition. The highest possible symmetry is that of an isotropic material whose properties are the same in all directions. In the case of β-brass the high-symmetry configuration corresponds to the disordered state at the higher temperature. In anisotropic materials whose properties vary in different directions, the symmetry is lower than for the isotropic case (see section 2.4). In the 3-pin soap-film model, when the movable pin C moves from inside to outside the arc ADB (figure 3.10), there is no overall change in the total symmetry of the complete soap-film system. Nevertheless, there is a change of local symmetry about the intersection point as the junction changes from type marked C (figure 3.10(a)) to type marked E (figure 3.10(b)). The symmetry at this point characterizes the pattern, so we use this symmetry to characterize the soap-film configuration. The lengths of the arms do not affect the local symmetry at the junction.

When the movable pin lies outside the arc ADB at C′, there exists threefold symmetry locally about E. It is equivalent to crystallographic symmetry 3m as additionally there is local mirror symmetry. When the moving pin lies inside the arc, there is no particular rotational symmetry. Crystallographers describe this as symmetry 1 as the pattern only repeats itself if it is rotated by 360°, i.e. 2π divided by unity. There is again local mirror symmetry which can be represented by m.

Order parameters are often used to represent changes of structure in a quantitative way. The order parameter is equivalent to the configuration parameter used previously. It is conventional that the higher the symmetry of a structure, the lower is its order parameter. Here, the higher value of the parameter should occur when the pin lies within the arc ADB and the lower value of the parameter when the pin lies outside the arc ADB. It is usual for the order parameter to go to zero in the higher symmetry (disordered) phase (see for instance Landau and Lifshitz 1980). An order parameter that gives a variation analogous to that for crystal phase transitions can be defined as

$$\eta = (A\widehat{X}B/60° - 2)^{1/2}$$

where $X = C$ for $x \leqslant x_0$, and $X = E$ for $x > x_0$.

Once again we are using $(\text{angle})^{1/2}$, just as we did to produce the energy versus configuration curves illustrated in figure 3.2. Also, the power of $1/2$ in the expression fits with phase transition theory. In addition it fits with order parameters and configuration parameters used in the next chapter. The variation of the order parameter η with pin distance x is shown in figure 3.16 and is demonstrated by program A.II.7 in Appendix II.

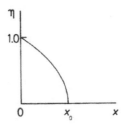

Figure 3.16. Variation of order parameter η with control variable x for the 3-pin model.

4

Soap-film models and catastrophe models

Elementary catastrophes as described by mathematicians are introduced. It is explained how the Zeeman catastrophe machine can be used to demonstrate the cusp catastrophe. Parallels are drawn between this machine and a four-pin soap-film model which can be described by a double-cusp catastrophe. Finally, the chapter returns to phase transitions and draws parallels between phase-transition theory and catastrophe theory.

4.1 INTRODUCTION TO CATASTROPHE THEORY

In the previous chapter we were preoccupied with sudden changes of state. In mathematics, these are called discontinuities. Catastrophe theory is about such discontinuities. It particularly concerns the singularities in our graphs where these discontinuities occur, and where it is not possible to use functions that change continuously. We have come to realize in earlier discussions that finding a mathematical form for the shapes of many of the soap-film structures in equilibrium is difficult enough. Certainly, we have not attempted to describe quantitatively how the shape of the film changes with time as it actually jumps between the different configurations.

Being a general approach to the problem of singularities and discontinuities, catastrophe theory is detached from the underlying mechanism of any specific discontinuity or jump. It is based on the description of our system by a finite number of variables, which we divide into two types. These two types are the **state** or **internal variables** and the **control** or **external variables** (also called **parameters**). How many configurations can occur depends on the number of control variables, and this is usually small even if there are many state variables. Provided that the number of control variables does not exceed

four, there are only seven possible types of catastrophes, called
the **elementary catastrophes**.

In the case of the soap film, the control variables must have
some relationship to the position of the moving pin or of the
moving section of framework that produces the changing film
configurations. The configurations or soap-film patterns can be
defined by stating the coordinates of the junctions between the
separate components of the soap film. When the film pattern
changes, the positions and even the number of junctions between
components change. Catastrophe theory predicts the qualitative
behaviour of the system on the basis of an underlying potential
energy equation.   This potential energy equation is able to
describe the overall energy of the soap film at equilibrium and
also the energy for all configurations close to equilibrium.

What we need to see is a relationship between the equilibrium
energy variation for the soap film and the equilibrium surface
underlying catastrophe theory as the control variables change.
We need to relate the control variables within catastrophe theory
to the controls within the soap-film model.

The two simplest elementary catastrophes are the fold and
the cusp (see for instance Saunders 1980, Poston and Stewart
1978). We will introduce both of these in order to develop the
arguments, but we shall see in due course that for our soap-film
models we need a double-cusp catastrophe model. The potential
function for the fold is defined by

$$V(x) = x^3 + ux$$

and that for the cusp by

$$V(x) = x^4 + ux^2 + vx.$$

u and v are control variables and x is a single state variable.
It is perhaps worth noting here that if we had an x^2 term as
well as the x term in the cubic expression for $V(x)$, then the
squared term could be eliminated by a change of the variable x.
Similarly in the case of the quartic expression for the cusp, an
x^3 term can be eliminated by change of variable. In both cases,
a new variable x' is used such that $x' = x + b$, where the value
of b is chosen to remove either the square or cubic term.

4.2   THE FOLD CATASTROPHE

The fold involves one control variable and one state variable.
The equilibrium surface, as defined by the equilibrium values for

the system, is obtained by the standard mathematical technique of differentiating the expression $V(x)$ for the potential energy, and equating the result to zero:

$$3x^2 + u = 0.$$

This equation gives x as a function of u. Note that the curve, produced by plotting x whilst varying the control parameter u, does not represent the variation of energy but the variation of x at equilibrium. Having obtained the values of x at equilibrium, the variation of energy at equilibrium can be calculated by inserting these values of x into the energy equation. The condition for bifurcation (i.e. the condition where we go over to two solutions for $V(x)$ as compared with the possibility of one solution) is:

$$6x = 0.$$

We obtain this by differentiating a second time. We can also identify this $x = 0$ condition by looking at the fold as plotted in figure 4.1. Thus, the bifurcation is the single point $u = 0$. This point divides the control space into two regions, $u > 0$ and $u < 0$. If $u < 0$, there are two critical points for $V(x)$, a minimum (stable equilibrium) and a maximum (unstable equilibrium). If $u > 0$, it turns out that there are no *real* solutions.

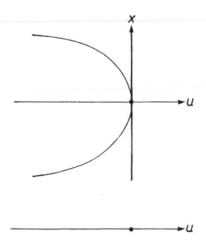

Figure 4.1. The fold catastrophe.

4.3 THE CUSP CATASTROPHE

The cusp (the single cusp) catastrophe model involves two control variables and one state variable. It is not strictly applicable to the two-dimensional soap-film patterns under consideration. The fact that this model has two control variables and one state variable means that it should correspond to a model where the control pin is free to move in two directions, but the junction is free to move in one dimension only. We shall need a model which allows movement of both the control pin and the film junction in two directions. Nevertheless, we need to understand the details of the cusp catastrophe in order to appreciate the arguments that follow.

The equilibrium surface for the cusp (but note once again that this is a surface for equilibrium values of x, not the equilibrium energy surface) can be found by differentiating the quartic function for $V(x)$ with respect to x and equating to zero, i.e.

$$4x^3 + 2ux + v = 0.$$

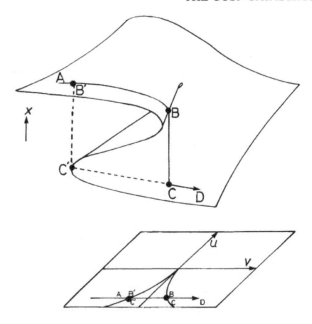

Figure 4.2 The equilibrium surface for the cusp.

This equilibrium surface is plotted using values of x as calculated by varying u and v. The surface is shown in figure 4.2. We see that the surface contains a very distinctive backwards and then forwards fold. As a consequence there are certain values of u and v where there are three values of x which fit the equation of the equilibrium surface. However, it turns out that one of the three values gives the condition for unstable equilibrium. The condition for singularity, beyond which there is only one solution, is found by differentiating once more with respect to x, and again equating to zero

$$12x^2 + 2u = 0.$$

Using the equations for the equilibrium surface and for the bifurcation condition as simultaneous equations, we eliminate x and obtain the relationship between u and v in the uv plane for the bifurcation. This is plotted below the equilibrium surface in figure 4.2. We see what happens if we move on the equilibrium surface from A through B and then from C to D. We have a smooth variation of x from A to B, a jump in x as we go to a new fold of the surface at C and then a smooth variation of x as we move to D. If we reverse our movement, we return to C', move upwards to a value of x at B' and back to A.

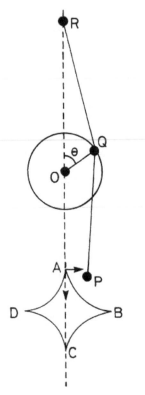

If we were to calculate the values of $V(x)$ for positions on the two paths we would find that although in one case x decreases in the jump and in the other case increases, the value of the energy $V(x)$ decreases for either jump. This is the origin of the word catastrophe, as the word was first used in French (the early theory was by René Thom), where it means a fall downwards. Notice that during passage along this path ABCDA, we do not move onto the middle leaf of the folded surface. This middle leaf corresponds to unstable equilibrium.

4.4   THE ZEEMAN CATASTROPHE MACHINE

As an example of a cusp catastrophe model, it is useful to consider the so-called Zeeman machine (Zeeman 1972). It is made from a thin cardboard disc mounted on a stiff board. The disc is pinned with a drawing pin to the board through its centre O (see figure 4.3) such that it can rotate. Two nearly identical rubber bands are fixed (pinned) to a single point Q on the disc near to its perimeter. One of these is stretched taut from Q and fixed to a point R on the base board. The other band QP is attached to a pointer at P with this pointer left free so that it can be moved around by hand.

Figure 4.3.   The Zeeman
catastrophe machine.

4.4.1   Operating the machine

Slowly, we move the pointer P around on the base board of the machine and watch what occurs. We find that as we move P there is rotation of the disc and hence movement of the point Q. However, there are positions on the base plane where a small movement of P causes a big jump in position of Q associated with a large rotation of the disc. Continued study of the motion indicates a diamond shape ABCD which marks the positions where these jumps occur. These are the lines marking the bifurcations.

Inside the diamond there are always two possible positions of Q for any one position of P. One position of Q is to the right of the dashed symmetry line and the other position of Q is to the left. It should be possible to find a third position where the wheel will just hold in place, but this is a position of unstable equilibrium, whereas the other two positions are entirely stable. Whether point Q lies to the right or left of the symmetry line depends on the previous path of the pointer P. If we move the pointer from left to right, the point Q will move from a negative angle to a positive angle (taking the vertical symmetry line as zero angle) as the pointer goes across the *right-hand edge* (edge ABC) of the diamond. If we then move the pointer from right to left, the angle will change from positive to negative as the pointer crosses the *left-hand edge* (edge ADC). Thus wherever the pointer lies within the diamond there is a possible negative and possible positive value of θ. Outside the diamond-shaped area, there is only one possible value of θ for any particular position of pointer P.

4.4.2 Discussion of what happens

An analysis of the Zeeman catastrophe machine is based on calculating the energy associated with stretching the rubber bands PQ and QR. The control variables are associated with the movement of pointer P to left and right (one parameter) and up and down (the second parameter). The analysis is carried out for small values of θ (so that one can use small angle expansions) and this restricts the region of validity to the neighbourhood of either the cusp point A or cusp point C. The state of the machine is specified by the single state variable θ, the angle made by line OQ with the axis of symmetry.

Although the control variables u and v finally used in the analysis are not the actual displacements to the pointer P, but are adjusted coordinates, and the state parameter x finally used is not the actual value θ, but an adjusted angle, we can see in principle without doing the mathematics how a cusp catastrophe equation is derived, involving one state variable and two control variables. The upper or lower corners of the diamond, A or C, represent the bifurcation cusps which we have identified already. The left-hand and right-hand corners of the diamond D and B, are also cusps, but they correspond to potentials where the quartic term is $-x^4$ rather than x^4. Hence, unstable equilibrium states exist here. These states at D and B are the reverse of the stable equilibrium states at the other corners A and C, and these two corners are so-called dual cusps.

Figure 4.4 shows the uv plane, which is related to the plane in which the pointer P moves. The shaded region is the region of bimodality. At any point in the uv plane we can draw the corresponding V–x curve. Each of these curves shows how the potential energy varies with angle θ (except that it is an adjusted angle x) for a particular position of the pointer. Each curve therefore shows the variation of potential energy as the disc rotates, passing from a non-equilibrium position, through the

---

Box 4.1   Properties of the cusp catastrophe

There are five properties typical of the cusp catastrophe (Zeeman 1976), and these are demonstrated by the Zeeman catastrophe machine:

(i) *Sudden jumps* occur.

(ii) There is *hysteresis*; i.e. as P moves back and forth across the diamond area the jumps do not occur at the same point when movement is to the right and when the movement is to the left.

(iii) There is *divergence*. If P starts very close to the axis but outside the diamond, whether Q is to the right or left depends on which side of the free end of B the pointer P passes. Two nearby trajectories produce very different behaviours.

(iv) *Bimodality*. There are two possible stable states when P exists within the diamond area.

(v) *Inaccessibility*. For P above or below the diamond area, the wheel can be set with any angle θ by moving P to left or right, but this is not possible when P is moved to left or right on a level with the diamond.

---

equilibrium position and then to to a non-equilibrium position on the other side.

4.4.3   Comparison with the earlier four-pin soap-film model

We see that the potential energy versus θ curves are similar to those already shown in figure 3.2 for the soap-film model having a fixed pair of pins plus a movable pair. This is provided we only consider the part of the plane where u is negative. Effectively, when we obtained these curves, we were looking at a cross-section of a cusp-type potential surface with u fixed. We did this by restricting consideration to a rectangular (but including square) arrangement of the pins and allowing the junction points of the film to move with one degree of freedom only. The configuration variable β for the soap-film model corresponds to θ in the Zeeman machine. The control variable x was varied by altering the length of the rectangle. A set of $V-\beta$ curves was

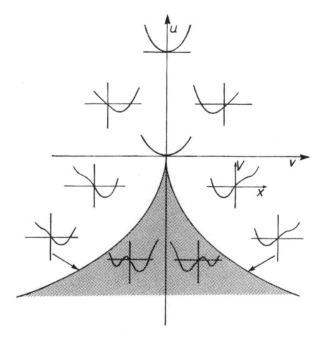

Figure 4.4   Variation of energy V with parameter x for differing values of the control parameters u and v for the Zeeman machine.

obtained for the soap-film model corresponding to a set of V–x curves for the Zeeman machine for a fixed value of u.

4.5   ILLUSTRATING DOUBLE-CUSP CATASTROPHES WITH SOAP FILMS

A particularly nice and alternative soap-film example is that of four pins, where three are fixed and only the fourth is movable. We could set three of these at three corners of a square. But a more interesting and highly symmetrical case can be considered in which three pins form the corners of an equilateral triangle and the fourth pin moves freely around in the plane. Because the film establishes a junction which can move around in two dimensions, unlike in the Zeeman model where the point Q can be defined by the single variable θ, we need a model which has two state variables. This involves a double-cusp catastrophe model, which has somewhat similar, but not identical, properties to the single-cusp catastrophe model.

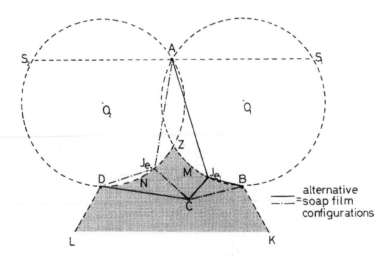

Figure 4.5 Double-cusp (soap film) catastrophe machine showing two equilibrium patterns.

4.5.1 Equilibrium film pattern in the 'three fixed pins plus movable pin' model

Figure 4.5 shows the two equilibrium film patterns which can be established when the control pin C is in an arbitrary position and A, B and D are the pins set at the vertices of the equilateral triangle. The shaded area delineates the region in which two configurations are possible. Arcs AMB and AND are those arcs of circles, centres O_1 and O_2 respectively, such that any angle $A\widehat{M}B$ and $A\widehat{M}C$ is 120°. Hence it is possible to have an equilibrium film junction J_e lying on either of the arcs. The two configurations are equivalent to those already shown in figure 3.1 for joining the four pins in a square formation. One configuration has equilibrium junction J_{e_1} lying on arc AMB and the other has an equilibrium junction J_{e_2} lying on arc AND. So one arrangement has film length $AJ_{e_1} + J_{e_1}B + J_{e_1}C + CD$ and the other arrangement has film length $AJ_{e_2} + J_{e_2}D + J_{e_2}C + CB$. Whichever pattern the film happens to form first, it cannot jump to the alternative pattern within the shaded area, even if this would reduce its length

4.5.2 Moving the control pin

The two arcs of the circles intersect at point Z, which is the circumcentre of the equilateral triangle DAB. Figure 4.6 sketches the changes of pattern which occur as the control pin is moved.

Figure 4.6(a)

When the control pin C is moved to Z, the two alternative film patterns become a single three-component pattern of total length $AZ + BZ + DZ$.

Figure 4.6(b)

If the control pin is moved a sufficient distance away from Z, it reaches a region where the film sub-divides from four components to five as in figure 3.1 for the square arrangement. This will occur should angles $J_{e_1}\widehat{C}D$ or $J_{e_2}\widehat{C}B$ become less than $120°$.

Figure 4.6(c)

However, we are only interested in what happens when C lies very close to point Z. Figure 4.5 already considerably exaggerates the displacement of C from Z compared with the displacement of interest in order to show the pattern clearly. From the discussion earlier (see page 18), we know that $AJ_{e_1} + J_{e_1}B + J_{e_1}C$ equals length CS_1, where ABS_1 is an equilateral triangle with all vertices lying on the circle with centre O_1. If we take the pin separation as unit length, the radius of the circle is $1/\sqrt{3}$. Hence the total length of the film at equilibrium is $DC + CS_1$, and CJ_{e_1} is coincident with part of line CS_1. Similarly the total length of the soap film in the alternative configuration is CS_2 (where S_2 is an equivalent point to S_1), and CJ_{e_2} is coincident with part of line CS_2.

Figure 4.6(d)

If the control pin C moves beyond the shaded area to left or right, only one equilibrium film pattern is possible. Suppose the film exists in a pattern with equilibrium junction J_{e_1} (drawn with continuous lines in figure 4.5) and we move the control pin towards and across arc ZND. The equilibrium junction J_{e_1} will move around the arc BMZ towards Z but no distinctive change of film pattern will occur even when C crosses the arc DNZ of the left-hand circle.

Figure 4.6(e)

Remaining with this film pattern, if we move C back towards arc BMZ, there will be a sudden change to the alternative configuration as C touches the arc BMZ, and as C converges with the equilibrium junction point J_{e_1}. Lengths AJ_{e_1} and DJ_{e_1} now meet up at C and move rapidly to the left to produce a new equilibrium junction J_{e_2} on arc DNZ. The film has moved to the alternative configuration. This transition can be demonstrated using *Mathematica* program A.II.8 in Appendix II. The hysteresis

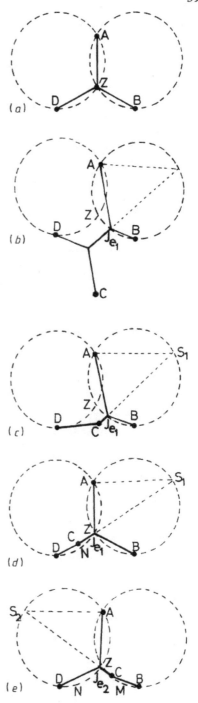

Figure 4.6. Moving a single control pin in the soap-film model.

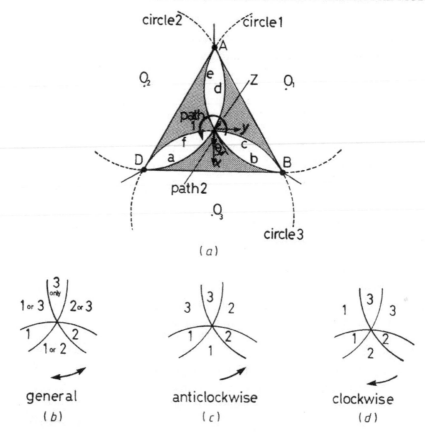

Figure 4.7 (a) Bifurcation pattern for complete freedom of movement of pin in the double-cusp catastrophe model. (b), (c) and (d): bifurcation boundaries for the control pin moving in different directions.

loop is illustrated in program A.II.9 where the rate of change of film length with movement is calculated and plotted.

The boundaries of the shaded region are bifurcation lines since they separate the plane in which C moves into regions where the model is monostable and a region where it is bistable. The shapes of the boundaries are partly circular arcs and partly straight lines, at least in ideal circumstances where there is no 'sticking' of the film. The straight line boundaries BK and DL make angles of 120° with line BD.

4.5.3 Moving the control pin 360° around the cusp point

A more complete picture emerges when the control pin is allowed complete freedom through 360° around Z, although staying close

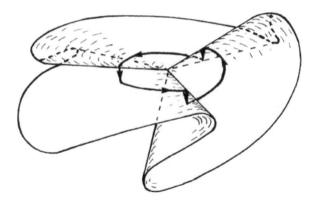

Figure 4.8 Complete equilibrium surface for soap-film model involving the movement of a control pin within the area defined by a triangular set of pins.

to Z. We certainly restrict the control pin within the equililateral triangle delineated by the other three pins. The result is shown in figure 4.7(a) where three areas are shaded. In each of these shaded regions two configurations are possible. The boundaries of the regions are described by arcs of circles, centres O_1, O_2 and O_3, which pass through pairs of points A and B, A and D, and B and D. The centres of the circles subtend angles of 120° to the pairs of pins. In the unshaded regions, only one soap-film configuration can occur.

By looking at the arrangement in figure 4.5, we can see that the control point can be joined directly to pin B and we will designate this configuration 1. Or it can be joined directly to pin D, designated configuration 2. As a third alternative, it can be joined directly to pin A when the configuration is designated 3. We mark up the possible configurations in each region and the possibilities are displayed in figure 4.7(b).

However, we have already discussed the fact that in a cusp catastrophe the configuration or energy state in which the system exists can depend on the path taken previously. If we consider a rotation of the pin around Z, the jumps occur at different angular positions according to whether we rotate in a clockwise or in an anticlockwise direction. Jumps in configuration will occur only at the boundaries between shaded and unshaded regions and such jumps will occur only if the film is not already in the appropriate configuration for the particular unshaded region. In figure 4.7(c) the configurations are shown for the anticlockwise rotation and in figure 4.7(d) they are shown for the clockwise rotation. Angles are measured relative to an x axis taken as $-ZA$.

In order to have three possible jumps in a complete rotation we have an equilibrium surface of the form shown in figure 4.8. It is as if we have taken three cusp catastrophe surfaces and joined the cusp regions together.† As with the single-cusp model, equilibrium values of the configuration parameter are then obtained for all points on this surface, from which values for the equilibrium potential energy are derived. Whether the pin rotates in either a clockwise or an anticlockwise direction, it is continuously climbing a potential energy gradient and then dropping down again. This is analogous to the artificial example of descending a continuous set of steps (figure 4.9). This illusion has been used very successfully in the picture by the Dutch artist M C Escher *Ascending and Descending* in which monks continually ascend or descend a set of steps. Escher has also used the technique in his lithograph *Waterfall* (figure 4.10) in which water continually moves around a series of channels and falls down a waterfall with uninterrupted flow. This is despite the water driving a waterwheel as it drops down the waterfall. The analogy is excellent. Whereas the two-dimensional picture is an illusion, the energy changes for the soap film are real. Any energy lost (as thermal energy) as the film is stretched must be supplied to the film as we move the pin.

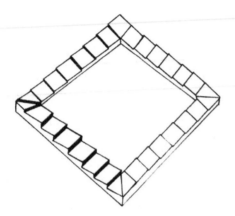

Figure 4.9. Continually descending (or ascending) steps.

Figure 4.11(a) shows the variation of length, and hence of potential energy V of the soap film, with angular position of the control pin. The potential energy axis is calibrated on the basis that the fixed pins have unit separation (with unit energy corresponding to unit film length) and that the pin moves in a circle very close to Z. The minimum possible energy corresponds to a total film length of magnitude $3R$, where $R = 1/\sqrt{3}$ is the radius of each of the three circles. $3R$ is the length of three arms meeting at Z. Calibration will be different if pin C is moved a significant distance away from Z, but the variation remains similar (provided the distance of C from Z is kept less than R).

The jumps correspond to first-order transitions. The rate of change of energy with angle, $dV/d\theta$, shows the regions of hysteresis as seen in figure 4.11(b). We can also demonstrate a second-order transition with the corresponding variation of

† We should not confuse the fact that three cusps are encountered as the control pin completes a 2π rotation with the fact that we need a double-cusp model to allow for the two degrees of freedom of the soap-film junction. The Zeeman catastrophe model exhibits four cusps of which two correspond to stable conditions and two to unstable conditions, even though operation of the machine can be represented by the single-cusp catastrophe model.

Figure 4.10 *Waterfall*, 1961. Lithograph by M C Escher. Collection Haags Gemeentemuseum - The Hague. © 1961 M C Escher Foundation - Baarn-Holland. All rights reserved.

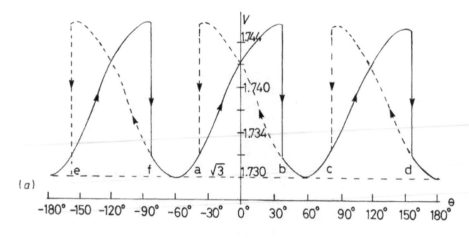

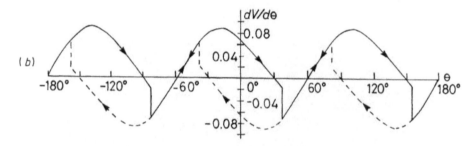

Figure 4.11 (a) Variation of soap-film energy V (proportional to length) with angle θ as the control pin rotates around the central point Z (path 1). (b) Rate of change of energy $dV/d\theta$ with angle θ.

energy, together with first and second differentials, by moving the control pin through Z in any *radial* direction; for instance we can move the pin along the x axis. As we would expect, there is no discontinuity either in the energy V or in its first differential dV/dx but there is in the gradient of the first differential; so the discontinuity shows up as a jump in the second differential d^2V/dx^2 at $x = 0$ (figure 4.12).

Further *Mathematica* programs are given in Appendix II to demonstrate this four-pin 60° model. Program A.II.10 shows how film length L varies with angle θ as the pin moves around Z and A.II.11 shows the variation of $dL/d\theta$ with θ, also as the pin circulates around Z.

4.5.4 Inclusion of non-equilibrium configurations

If we wish to represent the two-dimensional movement of the control pin C in full mathematical detail, we need to use two angles. (It proves mathematically easier to use two angles rather than an angle and a radial coordinate.) Everything we have demonstrated so far concerns the equilibrium configurations. A complete analysis appropriate to catastrophe theory needs additionally to consider non-equilibrium configurations. The theory should not be expected to apply when the control pin is distant from point Z, although the analysis fits well with the actual geometry even when the control pin is moved outwards from Z by a distance equivalent to say 20% of the fixed-pin separation. Mathematically we need angles which remain small as the pin moves away from Z. It is convenient to use points S_1 and S_2 (see figure 4.5 earlier) from which to measure the angles. Two corresponding configurations are shown in figures 4.13(a) and (b).

In the full analysis which does not assume an equilibrium configuration, the junction point is displaced by small angles relative to zero angle lines ZS_2 and ZS_1. We have called these angles λ and κ. As indicated already, the state parameters will be the square roots of the angles or in practice the square roots of angles adjusted by an additive component, which depends on the position of the control pin. Alternative angles to angles λ and κ can be used; the additive component will turn out to be different but the analysis is similar. Without going into full detail it suffices to say that a potential of the form

$$V = \mathbf{A}P^4 + \mathbf{B}Q^4 + \mathbf{C}P^2Q^2 + \mathbf{D}P^2 + \mathbf{E}Q^2$$

can be obtained, in which $\mathbf{A}$, $\mathbf{B}$, $\mathbf{C}$, $\mathbf{D}$ and $\mathbf{E}$ are control variables related to the geometry of the three fixed pins and the position of the control pin C, with their dependence on the position of the control pin C being quite complicated, and being obtained by assuming small angle expansions. P and Q are state variables related to, but not equal to, the angles κ and λ. The small angle approximations set the numerical limits to the model.

As we might expect from the way we have set up the mathematics and from our knowledge of the physics of the model, the $P = 0$, $Q = 0$ solution corresponds to the high-symmetry phase when the control point and junction coincide at Z. If we assume either P zero or Q zero, we have the single-cusp equation but without a linear term. Hence, we can see that varying either angle λ or κ (and keeping the alternative angle constant) will give us potential energy plots as functions of $\lambda^{1/2}$ or $\kappa^{1/2}$ of similar forms to those shown for the cusp catastrophe model in figure 4.4 but always with $u < 0$.

The minimum (or minima) in these plots will give the equilibrium position (or positions) of the junction point and

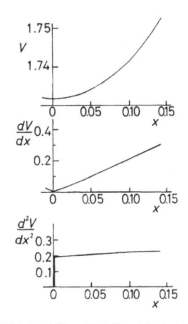

Figure 4.12. Variation of energy V with distance x as the control pin moves along path 2. Also dV/dx and d^2V/dx^2.

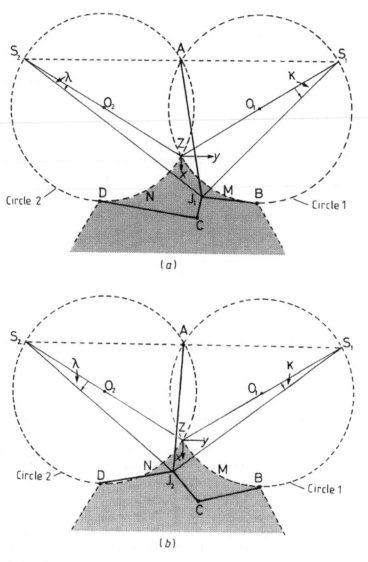

Figure 4.13 Soap-film pattern for the double-cusp model when the film junction is not in an equilibrium position. (a) and (b) show the two alternative configurations.

the corresponding values of $\lambda^{1/2}$ and $\kappa^{1/2}$. These equilibrium positions for the junction point $(J = J_e)$ must lie on CS_2 or CS_1 (or an equivalent CS_3). They also lie on the arcs of the circles 1, 2 and 3 shown in figures 4.7(a) and 4.13. Thus we have the equilibrium positions defined for any fixed control point C.

Box 4.2   The double-cusp model

Mathematically, this has an equation of the form

$$V = \mathbf{A}P^4 + \mathbf{B}Q^4 + \mathbf{C}P^2Q^2 + \mathbf{D}P^2 + \mathbf{E}Q^2$$

as indicated for the current soap-film model. A detailed analysis is beyond the scope of this book but we can see that we can partially differentiate the equation with respect to P and Q to obtain

$$\partial V/\partial P = P(4\mathbf{A}P^2 + 2\mathbf{C}Q^2 + 2\mathbf{D}) = 0$$

and

$$\partial V/\partial Q = Q(4\mathbf{B}Q^2 + 2\mathbf{C}P^2 + 2\mathbf{E}) = 0.$$

The solutions are:

$$P^2 = (\mathbf{CE} - 2\mathbf{BD})/(4\mathbf{AB} - \mathbf{C}^2)$$

and

$$Q^2 = (\mathbf{CD} - 2\mathbf{AE})/(4\mathbf{AB} - \mathbf{C}^2).$$

This solution gives the bifurcation lines which, in the xy plane of the soap-film model, are the circles through pins A, B and C. The energy jumps are as illustrated in figure 4.11.

This argument fits with the energy changes already identified as we rotate C around Z. For a brief mathematical discussion see box 4.2.

4.6 CHAPTER APPENDIX; PHASE TRANSITIONS AND LANDAU THEORY

4.6.1 Landau theory of second-order phase transitions; the order parameter

Second-order phase transitions are best understood on the basis of an order parameter η. This describes a change of physical property, such as a change of structure, when a system passes

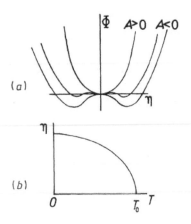

Figure 4.14. Second-order phase transition: (a) variation of thermodynamic potential Φ with order parameter η for $A > 0$ and $A < 0$; (b) temperature dependence of the order parameter η.

property, such as a change of structure, when a system passes through a phase transition (Landau and Lifshitz 1980, Izyumov and Syromyatnikov 1990). As indicated already, the parameter must be non-zero in the non-symmetrical phase and zero in the symmetrical phase. In the neighbourhood of the transition, the thermodynamic potential, which may be one of a number of alternatives, including Gibbs free energy and Helmholtz free energy, according to the experimental conditions, can be expanded in a power series of η:

$$\Phi(P, T, \eta) = \Phi_0 + A\eta^2 + C\eta^3 + B\eta^4 + \cdots$$

where A, B, C are functions of P and T, following the notation of Landau. The expansion will not hold within a very narrow interval involving the transition itself. There is no term in η to first order if the states have different symmetry either side of the transition.

We can see how Φ varies for $A > 0$ and for $A < 0$ with $C = 0$ (figure 4.14(a)). For Φ to be a minimum on the high-symmetry side of the phase transition (where three components of the soap film meet at 120°), A is greater than 0. For Φ to be a minimum on the low-symmetry side, A is less than 0. As the sign of A changes at the change of state, A must be zero at the phase change.

In the vicinity of the phase transition, assuming the transition occurs at temperature T_0, the coefficient for the quadratic term in η can be assumed to be a linear function of displacement temperature measured from the transition temperature which takes the form

$$A = A'(T - T_0).$$

If Φ, as a function of η, is a minimum for $A > 0$ at $\eta = 0$ (figure 4.14(a)), the third-order term will be zero ($C = 0$) and the fourth-order term should be positive ($B > 0$). Using $\partial\Phi/\partial z = 0$ to give minimization leads to

$$\eta(A + 2B\eta^2) = 0$$

$$\eta = 0 \qquad \text{or} \qquad \eta^2 = -\frac{A}{2B} = \frac{A'(T_0 - T)}{2B}$$

when

$$\eta \propto (T_0 - T)^{1/2}.$$

This power dependence is significant. We use similar arguments with other thermodynamic quantities. In our soap-film models we have used variable length or variable angle parameters and found that we must use the square root of an angle as an order parameter. Figure 4.14(b) illustrates the temperature dependence of the order parameter and this can be compared with figure 3.16 on page 49 which showed the order parameter dependence on the control parameter x for the soap film.

4.6.2 Applying Landau theory to first-order transitions

For the analysis in the previous section there was no cubic term in η as C was taken as zero. If C is finite, the transition becomes first order. Using the condition $\partial\Phi/\partial z = 0$ gives

$$\eta(2A + 3C\eta + 4B\eta^2) = 0$$

which can be solved to produce two non-zero solutions

$$\eta_\pm = -\frac{3C}{8B} \pm \left[\left(\frac{3C}{8B}\right)^2 - \frac{A}{2B}\right]^{1/2}.$$

These solutions are real when $A \leqslant 9C^2/32B$.

We can plot η as a function of A. Once again we can relate A to a change of temperature $(T - T_0)$, in which case we can show the variation of η with temperature (figure 4.15). We see that the curves for η_+ and η_- are the same (merely the sign reversed) so that we need consider only one curve as a solution. This is because the equation of state for our system is invariant to changing C from positive to negative at the same time as changing η from positive to negative. Between the marked points T_1 and T_2 there are two non-zero values of η_+ (or alternatively η_-) and so there are two possible values of the order parameter within this region. This suggests that there can be partial disorder existing between the total disorder of one phase (say a gaseous phase) and the much greater order of the other phase (say liquid phase) and in particular hysteresis can occur in changing from one phase to the other. For $A > A_0 = 9C^2/32B$, a non-zero value of η is impossible and this defines the absolute instability boundary of the disordered phase. Temperature T_2 is the maximum superheating temperature of the liquid and temperature T_1 is the minimum supercooling temperature of the gas. As we know, once a system starts to change phase from a superheated or supercooled state, the change usually occurs quickly, although not in all phase transitions. Nevertheless, statistically we can consider a partially ordered phase. A true first-order phase transition can occur anywhere within the temperature region $T_1 < T_c < T_2$.

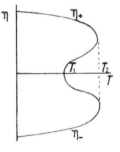

Figure 4.15. First-order phase transition; order parameter η plotted as a function of T.

4.6.3   Two-dimensional order parameter and the addition of an external field

We have expressed the coefficient A in the equation for the thermodynamic function as a linear function of temperature. But originally we portrayed A as a function of pressure P in addition to being a function of T and the order parameter η. We could carry out a similar exercise with respect to P as we did with respect to T. If we expand our thermodynamic function allowing for both temperature and pressure it would involve a two-dimensional order parameter, i.e. an order parameter with two components. Expansion in this way is analogous to allowing the film junction in one of the soap-film models to move around in a two-dimensional plane. It is not surprising therefore that this involves more complicated equations. The characteristic equation for the potential for a system showing phase transitions due to a two-component order parameter (components η_1 and η_2) takes the form (Izyumov and Syromyatnikov 1990, p 168)

$$\Phi = A(\eta_1^2 + \eta_2^2) + B(\eta_1^4 + \eta_2^4) + C\eta_1^2\eta_2^2$$

which is analogous to the equation for a symmetric double-cusp model. Although phase transition theory has somewhat different starting points from those of catastrophe theory, the conclusions must, in the end, coincide.

In addition, we can apply a field to our thermodynamic system; this field can be electric, magnetic or gravitational. To do this we need to incorporate in the expression for thermodynamic potential Φ an extra energy term resulting from the application of the field. This term can be considered as proportional both to the magnitude of the field and to the order parameter. Thus, the further term within the expression for Φ is a linear term in η, whereas in the previous two sections of this chapter a linear term was omitted in the expansion for Φ.

The soap-film models used to demonstrate phase transitions have been assumed to be held horizontally and hence there is no effect due to the gravitational field. Should such a field be applied, by say setting the models vertically, an analogous situation will be established. Indeed we shall see in the next chapter that the stability of a film established within a wedge differs according to whether the perspex wedge is held horizontally or vertically in the gravitational field.

In conclusion, we can consider the thermodynamic potential to be in the form of an expansion of terms in η (the order parameter); we realize that we can have linear, square and quadratic terms in η; and we see that we can have a two-dimensional (or even a three-dimensional) order parameter. Hence, a large range of possibilities exist and there are various mathematical expressions to represent the alternative forms of phase transition.

On the one hand we have the thermodynamic potentials expressed as functions of order parameters to demonstrate phase transitions, usually involving hysteresis. On the other we have the potentials in catastrophe theory expressed as functions of state to demonstrate the existence of bifurcations at the points where a monostable system becomes bistable. The two are equivalent.

The coefficients (the control variables) in the mathematical expansions for the catastrophe models have complicated dependencies on the position, as expressed in Cartesian coordinates or angular coordinates, of the control pin. It becomes clear why the parameters in the soap-film models are the square root of an angular displacement. They are equivalent to the square root of a temperature variation or of a pressure variation. It is also clear why the coefficients of the linear, quadratic, etc. terms, depend on the position (in Cartesian or angular coordinates) of the control pin. They are similar to those in a catastrophe model, and to those in expansions for thermodynamic potentials involving temperature and pressure.

5

Film within a wedge—the catenoidal surface

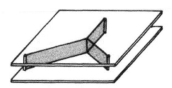

Figure 2.1. *Revisited.* Two-dimensional film pattern established between three pins arranged in a triangle.

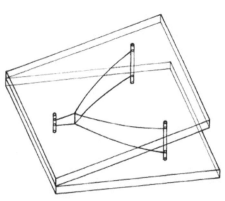

Figure 5.1. Soap-film pattern established between three pins within non-parallel plates.

Although it is difficult in general to describe soap film surfaces mathematically, the catenoidal surface can be so described, and has interesting properties. The film established within a wedge is catenoidal in form and is relatively stable. Therefore, it is used both as an example of such a surface and as a method of establishing a film of variable thickness for demonstrating thin-film interference fringes.

5.1 FILM BETWEEN THREE PINS JOINING NON-PARALLEL PLATES

On page 13 we discussed the film pattern which is set up between three pins arranged at the corners of an equilateral triangle and mentioned that if the plates are not parallel, the film pattern changes in an interesting way. Suppose that the plates touch along one edge such that the space between is wedge-shaped. The obvious arrangement of pins is that in which two of the pins are equi-distance from the apex of the wedge. The pattern changes from that shown previously in figure 2.1 to that shown in figure 5.1. What takes us by considerable surprise is that the arm in the narrow region of the wedge is *shorter*. We might think that the arm within the narrow portion of the wedge should be longer because we would expect less area of film to be needed here. Closer inspection of the film shows that the other two arms are distinctly curved. They are curving into the narrow portion of the wedge so as to keep their areas small. This is the dominant effect and this feature leads to the third arm being considerably shortened.

5.2 FILM WITHIN A WEDGE

The next step is to construct a wedge involving the two pins

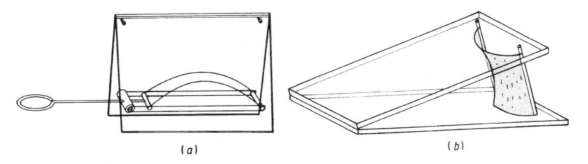

Figure 5.2   (a) Wedge model with soap film. (b) Side view of the film.

equidistant from the apex of the wedge and to omit the nearer
pin. In order to see more clearly what is happening, it is useful to
allow variable separation of the pair of pins. To achieve this, one
pin is fixed and the other pin, attached to a handle, moves back
and forth in grooves within the perspex plates (figure 5.2(a)).

The open side of the wedge is dipped into soap solution
so that a single-component soap film is established between the
two pins. We find that it bends into the wedge in a curve whose
shape we might recognise as similar to that of a hanging chain
suspended between supports. The curve is a catenary and the
film is part of a catenoid. The film may tend to align along the
grooves of the perspex plates but it can be blown easily into its
equilibrium catenoidal form. Provided the separation of the pins
is less than a critical value, the film is stable.

Looking at this film we might think the overall forces within
the film are unbalanced, but this cannot be so. The curvature we
have identified as being associated with the shape of a catenary
must be balanced by a curvature in a perpendicular plane. This
second curvature is immediately obvious if we look along the
length of the wedge (figure 5.2(b)). The relationship between
principal radii for a soap film curved in three dimensions has
been discussed already (page 10). The film is a section of a
catenoid of revolution about an axis which is the vertex axis of
the wedge. This becomes clearer by looking at figure 5.3. It is
similar to cutting a wedge-shaped slice from a circular cake. A
wedge angle of about 15° is suitable. It is possible to use any
size of angle, except that at large angles it becomes important
that the pins take on an appropriate curvature, rather than being
straight. This curvature enables the pins to meet the perspex
plates at 90°.

We now move the variable pin.   What is particularly
interesting is that the soap film moves towards the vertex or back
outwards (i.e. the film movement has a component perpendicular

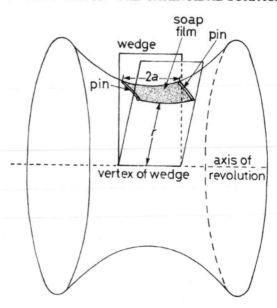

Figure 5.3 Complete catenoid of which the soap film within a wedge is part.

to the pin movement), according to whether the separation is increased or decreased. The film is highly stable and responds to the pin movement, although, if the pin moves rapidly, the film lags. If the pin separation is increased beyond a certain critical value, which turns out to be greater than 1.33 times the perpendicular distance of the pins from the wedge vertex (see the mathematical appendix to this chapter), the film moves off in a distorted catenoidal form to meet the wedge vertex. Once the centre of the film meets this vertex, the film separates into two triangularly shaped films, each bounded by a pin plus the surfaces of the wedge. If the variable pin is moved inwards after the film movement has started, we find that there is a position prior to which the film can be brought back to an equilibrium catenoidal shape. Beyond this film position, no amount of pin movement will prevent the film travelling onwards to its triangular shape. The film has moved beyond a position of unstable equilibrium, and is descending an energy slope.

It is widely known that the soap film formed between two parallel wire hoops (or rings) of equal radius has a catenoidal surface. The relationship between this film and the restricted film within the wedge is obvious from figure 5.3. As the separation of the hoops is increased, equivalent to increasing the pin separation within the wedge, the soap film gradually necks in. A similar instability occurs to that shown within the wedge; the soap film splits into two flat circular surfaces stretched across the hoops. This instability and the switch to the circular surfaces occurs remarkably rapidly. As a consequence, it is much easier to study the instability within a wedge. The film lasts longer within the wedge, where there is reduced evaporation, and the motion of the film during the instability is slower.

It is clear that the film must be symmetrical about the axis joining the centres of the two coaxial hoops. It can be shown that there is a catenoidal solution of the form

$$r = c \cosh[a/c]$$

where the separation between the hoops is $2a$, c is a constant and r is the radius of each hoop. In the case of the wedge, $2a$ is the pin separation and r is the perpendicular distance of the pins from the vertex. A derivation of the catenoidal expression is outlined in the mathematical appendix (section 5.10). This solution for a stable state only applies provided $a < 0.663r$ (i.e. a hoop or pin separation less than $1.33r$).

5.3   SETTING THE WEDGE APEX UPWARDS IN A GRAVITATIONAL FIELD

In discussing the formation of a soap film within a wedge it has been assumed that the apex of the wedge and one surface of the wedge have been laid horizontally. In this way, the effect of gravity is not significant. The gravitational force does not affect the catenoidal curvature in the horizontal plane, although it may distort the curvature in a vertical plane. An estimate of the distortion can be obtained by looking at the interference fringes. These demonstrate a tapering of the film, thinner in the upper regions and thicker in the lower regions. If we mount the wedge with its open end sitting horizontally on the ground and the catenoidal surface arching between the two pins, rather like an arched bridge, then the effect of gravity will be to pull the film downwards from its original equilibrium position. This distortion arises because the film will have a mass associated with it, albeit small, and the gravitational force on this mass produces a measurable effect. This is sufficiently large such that even if the separation of the pins is slightly greater than 1.33 times the perpendicular distance of the pins from the apex, the film will remain in an approximately stable position.

However, the situation is not as stable and unchanging as might first be thought. There is drainage of water towards the pins due to gravity. The film re-establishes a new equilibrium as it becomes thicker close to the pins and thinner at the top of the curved surface. The film is sufficiently thin over most of its length to produce very attractive interference patterns when illuminated with white light. Sunlight or room lighting is quite sufficient for this. Such patterning is illustrated in the frontispiece (figure F1), which nevertheless fails to do full justice to the beautiful and dynamic colour distributions which are observed. The film drains such that a silvery or black coloration is observed in the top region of the curved surface. This corresponds to a film thickness of 150 nm or less. The film may even be only a couple of molecules thick, i.e. as little as 6 nm in thickness.

Many processes are operating simultaneously including convection, evaporation and suction, as a consequence of the variation of pressure and of concentration gradients. The equilibrium shape of the film is established relatively quickly, although even this process is slowed within the wedge compared with the establishment of film equilibrium between hoops. Draining is a much slower process. What is surprising is the very dynamic movement, including vorticity, which is set up in the film. At one moment, a region of the film may look silvery or black; a few seconds later it may have changed in colour to magenta or violet as more water molecules move into the region.

As the processes of evaporation and drainage proceed, the mass of the central region of the curved film becomes less and the film rises further into the narrow region of the wedge. Dry surfaces to the wedge can impede movement, so it is helpful to pre-wet the wedge. The film tries always to maintain equilibrium. Because it becomes extremely thin, it usually bursts rather than reaching that point where it can rapidly move upwards and separate into the two separate triangular regions.

5.4 INTERFERENCE COLOURS

Setting up a soap film in an upturned wedge is an especially good way in which to see interference colours. As indicated, the interference of white light within the film produces a range of interference colours which are characteristic of the film thickness. In the next section we look at the mathematical basis for the interference and how to calculate which colours will be observed.

Table 5.1 Interference colours produced by white light.

Colour	Thickness (nm)	Colour	Thickness (nm)
First order $p = 0$		Third order $p = 2$	
black	6–12	purple	396
silvery white	150	blue	410
(amber)	—	blue	428
magenta	201	emerald green	466
		yellow green	502
Second order $p = 1$		carmine	542
violet	216	bluish red	578
blue	250		
green	290		
yellow	322		
orange	348		
crimson	371		

Table 5.1 gives the first three orders of colours for various soap-film thicknesses up to a thickness of approximately 580 nm; this thickness is of comparable magnitude to the wavelengths of the components of white light as listed in table 5.2 below. This table of colours is taken from Lawrence (1929), where additional colours can be found listed up to eighth order. However, the first three orders cover the colours of interest here. The table assumes that the light is incident normally onto the film and a refractive index n of 1.41 is applicable. The presence of the soap or detergent increases the refractive index above its magnitude of 1.33 for pure water.

Within a relatively short time of being set up, the top portion of the catenoidal film exhibits a silvery-white coloration with strands of black or larger areas of black moving within the silver. Because of an optical phase change of π (i.e. $\lambda/2$) which occurs when light is reflected from the air–film interface, but which does not occur with reflection from the film–air interface, black corresponds to a film thickness approaching zero. Six to twelve nanometres, a thickness which gives rise to blackness, corresponds to little more than two organic molecules lined up back to back across the film. The name *Newton black film* is given to such a thin film.

The rapid changes of thickness, which occur within the black and silvery-white regions, arise because the film has thickness comparable with the range of the van der Waals interaction between the water molecules. Convection produces rapid colour variations, including at times regions of very highly coloured

Table 5.2 Approximate wavelengths (in air) for different colours.

Colour	Wavelength (nm)
red	680
orange	590
yellow	580
green	530
blue	470
violet	405

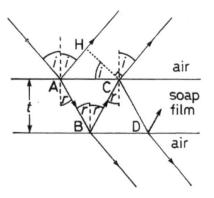

Figure 5.4. Reflection and interference of light from a thin film.

film. This phenomenon is discussed by various authors including Lawrence (1929) and Isenberg (1978).

5.5 THEORY OF THIN-FILM INTERFERENCE

Figure 5.4 shows the situation for monochromatic light of wavelength λ incident at an angle i onto a soap film of thickness t. Some of the light is reflected as it meets the first surface at A and the remainder is refracted into the film. The angle of refraction is r. The transmitted light is partially reflected at B and some emerges on the far side of the film in a direction parallel to the incident ray. The light which is reflected at B returns to meet the first surface at C, where some emerges at an angle of refraction equal to the original angle of incidence i, and some is further reflected internally.

The light reflected from the film at A and the light emerging at C will set up a reflection interference pattern. Rays emerging on the far side at B and D will set up a complementary transmission interference pattern. This is so provided the merging parallel rays in each case are focused by the eye of the observer. In practice, there will be further reflections from the two film surfaces but, as the reflection coefficient for soap films is not especially large (we would need silvered surfaces before we would have to take into account multiple reflections!), the approximation is reasonably good.

Let us consider rays emerging at A and C. The path difference is AB + BC − AH, where CH is perpendicular to the direction of the emerging rays. If we assume a refractive index of n for the film and 1 for air, the *optical path difference* (OPD) is

$$n(\text{AB} + \text{BC}) - \text{AH} = \frac{2nt}{\cos r} - \text{AC} \sin i.$$

Since

$$\sin i = n \sin r$$

(the well-known Snell's law but see also page 116) and

$$AC = 2t \tan r$$

then

$$\text{OPD} = \frac{2nt}{\cos r} - \frac{2nt \sin^2 r}{\cos r} = 2nt \cos r.$$

For reflections occurring at an interface which separates a less dense from a more dense medium, in that order, there is a phase change of π, giving an optical path difference of $\lambda/2$. This phase change will occur for reflection at A. For constructive interference, we require the light waves from A and C to be in phase, i.e.

$$2nt \cos r = (p + \tfrac{1}{2})\lambda$$

where p is zero or an integer.

Thus, constructive interference will occur at a fixed value of λ for a range of values of r or alternatively for a range of values of t. The intensity of the light will maximize and minimize as either r or t varies. We can go on to calculate the variation of amplitude, and hence of intensity, of the reflected light. We obtain for the reflected intensity a dependence of the form

$$I_{\text{reflected}} = 4 I_{\text{incident}} \mathcal{R} \sin^2 \left(\frac{2\pi}{\lambda} nt \cos i \right)$$

where I_{incident} is the intensity of the incident light and $\mathcal{R}$ is the fraction of the light reflected by the surface of the film at A. Thus, we can see how the intensity will vary as the incident angle of the light varies, or the film thickness varies or both. The angles at which constructive and destructive interference occur depend critically on the wavelength λ. If $I_{\text{transmitted}}$ is the intensity of the transmitted light, the value of the transmitted intensity can be found from

$$I_{\text{transmitted}} = I_{\text{incident}} - I_{\text{reflected}}.$$

With films illuminated by white light (such as daylight), the observed coloured fringes arise from the combination of a large range of frequencies. As indicated already, the light reflected off a very thin film will give the appearance of blackness because of the π phase change for reflection at the top surface of the film. In practice, the film will not usually be quite so thin as to appear totally black. As the film gets thicker we might expect a bright band corresponding to a thickness of $\lambda/(4n)$ or approximately 100 nm.† In practice, one sees a grey and then silvery-white coloration, probably because of the extended size of the source of illumination.

Slightly thicker regions of film (around 150 nm thickness) may look bright golden. As the blue gets cut out, the film goes through amber to magenta. At this point the green is at its second minimum corresponding to a thickness of $t = (1 \times 530)/(2 \times 1.41)$ nm $= 196$ nm where $n = 1.41$ (see page 77). Red reaches its second minimum at about 250 nm by which time blue has reached its second maximum, $t = (3 \times 470)/(4 \times 1.41)$ nm $= 250$ nm. This gives the film a bright blue colour. Between the magenta and the blue regions one should be able to distinguish a small region of violet. As the thickness of the film increases further (as one looks further down the curve of the catenary) one may be able to distinguish a yellow tinge to the blue, followed by green (290 nm), yellow (340 nm) and crimson (370 nm) before the film becomes purple (400 nm) and then blue again.

It should be noted that although blackness is produced by a Newton black film at a thickness slightly greater than 5 nm, one can sometimes identify a stable black film at the greater thickness of around 30 nm. This situation is sometimes referred to as the *common black film*. The blackness here arises when the film consists of multiple layers of the surfactant molecules existing in a repeated 'tail-to-head-to-head-to-tail' arrangement. Because

---

† The smallest thickness for which light is constructively reflected is given by

$$2nt \cos r = \lambda/2.$$

Assuming that light is incident perpendicular to the surface of the film and the angle of refraction r is $0°$, this gives the first condition for constructive interference as

$$t = \lambda/(4n)$$

whereas the second condition for constructive interference is $t = 3\lambda/(4n)$.

Destructive interference first occurs at a thickness approaching zero. The second condition for destructive interference is for

$$2nt \cos r = \lambda \quad \text{giving} \quad t = \lambda/(2n)$$

and the next condition for destructive interference is

$$2nt \cos r = 2\lambda \quad \text{giving} \quad t = \lambda/n.$$

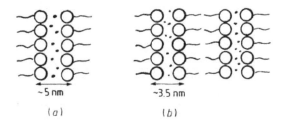

Figure 5.5 Arrangement of surfactant molecules in (a) a Newton black film and (b) a common black film.

the film is built up of repeat layers of the surfactants, light can be reflected as if from a film with narrower spacing. Figure 5.5 shows the head-to-tail arrangement of two layers of surfactant molecules forming a Newton black film and also the simplest doubling up of layers which can occur to form the common black film. This latter molecular arrangement in the film is analogous to the structure of certain Langmuir–Blodgett-type films which are discussed in Chapter 9.

5.6 FILMS BETWEEN TWO CONES

The catenoidal soap film can be set up between hoops of unequal radii. No longer can symmetry be assumed about the midpoint between the hoops. The same situation would occur if we set the pins within the wedge at unequal distances from the wedge apex. An interesting experimental variation on the problem is to use two funnels of different sizes (figure 5.6). One circular hoop is the outer perimeter of the smaller funnel. The soap film always meets the surface of the larger funnel at right angles. The point where this happens and hence the size of the second hoop is fixed for a particular funnel separation. As funnel separation is increased, a point is eventually reached when the film becomes unstable (Sinclair 1907).

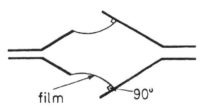

Figure 5.6. Film established between two funnels.

5.7 THE *H*-SURFACES OF REVOLUTION

Soap-film surfaces have a constant mean curvature (see page 10). This mean curvature is zero for any film with equal pressure on either side (i.e. for any film of minimal surface area). However, it is finite for bubbles with excess pressure on one of its two sides. Some of these surfaces are also rotationally symmetric and

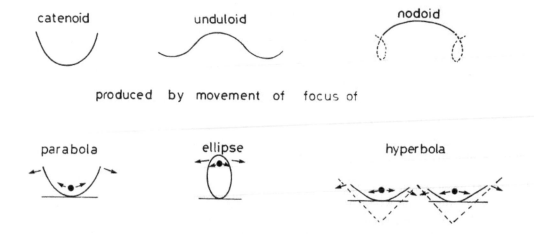

Figure 5.7   Formation mechanism for the catenoid, unduloid and nodoid.

are called *H*-surfaces of revolution by mathematicians. There are six such individually identifiable surfaces and the original delineation of these is associated, as with so many aspects of soap films, with Plateau.

The six *H*-surfaces of revolution are as follows. Firstly, there is the plane which is represented by a disc. Secondly, there is the catenoid. Both the disc and the catenoid have a mean curvature which is zero and the air pressure is equal on either side. Besides these two, there are four surfaces which do not have a mean curvature which is zero. These are the sphere, the cylinder, the unduloid and the nodoid. The first two are well-known, but the unduloid and the nodoid are less so. Geometrically, the unduloid can be formed from the loci (paths) of either of the foci of an ellipse, as the ellipse rolls about its axis on a flat surface. The unduloid is then the complete surface obtained by rotating this curve about the central axis. If the foci are close together (when the ellipse represents only a small distortion from a circle), the undulatory curve is only slightly wavy. Clearly the limiting case produces a straight line, and thus gives a cylindrical surface after rotation of the line. It is interesting to note that the catenoid can be formed by obtaining the locus of the focus of a parabola as it also rolls. The formation of the nodoid via the locus of one of the foci of a hyperbola is more difficult to obtain but the soap film will only take up part of this curve. Formation of the catenoid, unduloid and nodoid is shown in figure 5.7. For a fuller discussion the reader is referred, for instance, to Hildebrandt and Tromba (1985, p 162).

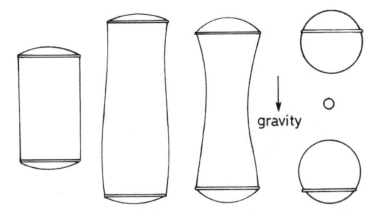

gravity

Figure 5.8 Film surfaces formed by blowing bubbles constrained by circular hoops.

5.8 ALTERING PRESSURE ON ONE SIDE OF A FILM

We have discussed altering the form of a soap film by changing the separation of the pins within the wedge or by altering the separation of the hoops in the case of the full catenoidal model. Once the separation is beyond a certain value, we have seen that the catenoidal film becomes unstable and we cannot bring it back. However, we might consider altering the pressure on one side of the film and this can be achieved by using end plates, either two equal circular plates fixed such that the line joining their centres is perpendicular to both plates, or using plates which are sectors of circles so that they block off the ends of the wedge. A valve arrangement through one of the plates is required so that pressure within the catenoid can be controlled. Such an experimental set-up has been used by a number of researchers (see for instance Erle *et al* 1970). The film now takes on one of three possible forms, all of which are *H*-surfaces of revolution, the catenoid, the unduloid or the nodoid.

Alternatively, films with these forms of curvature can be obtained by blowing a bubble which is restrained by two circular hoops (figure 5.8). The curvature at any point between the hoops will be determined by the excess pressure arising from the amount of air which happens to have been trapped within the bubble. Air can be added or extracted by inserting a wetted straw and blowing or sucking. Pressures and curvatures can be altered at will.

Thus by adjusting the air pressure, it is possible to give the film a cylindrical shape. Curvature will be in one plane only. This is possible because of the internal excess pressure and is

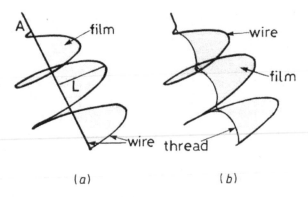

(a)   (b)

Figure 5.9 (a) Soap film formed within a wire spiral plus central wire axis combination—the helicoid. (b) Soap film within wire spiral plus thread.

unlike the situation for a spherical bubble. So the excess pressure will be given by

$$P = \frac{2\gamma}{R}$$

where R is the radius of the hoops and there are two surfaces to the film. This is obtained from the Laplace–Young equation on page 10. The ends of the film, surrounded by the hoops, curve symmetrically outwards as parts of spherical surfaces. The radius of curvature of these surfaces will be

$$P = \frac{4\gamma}{R}$$

so that here the radius of curvature in each direction is double that for the cylindrical part of the film.

By adding air, the pressure can be increased to produce a bulging film between the hoops. Alternatively, extracting air produces a decrease in curvature of the end surfaces and the cylindrical section curves in. As we would expect, it is only possible to produce a catenoidal shape for this part of the film if the hoops are sufficiently close. More usually, the film forms a 'waist' with minimum diameter occurring at half height. Eventually the necking leads to the film breaking and forming two separated spherical bubbles at the hoops (with a small free-falling bubble formed between).

5.9 FILMS WITHIN A HELIX—THE HELICOID

We can establish a soap film between a helical wire and a straight length of wire running down the central axis of the helix (figure 5.9(a)). Such a surface is called a helicoid; it is a surface which has been used in the design of spiral staircases. The minimal surface can be obtained by deforming a catenoid to which it is directly related. The catenoid is 'sliced' longitudinally between its two circular ends, and then, whilst one of the cut sides is kept fixed, the other is pulled upwards and rotated to produce the helicoidal surface. It is not surprising that the helicoid has been known from early mathematical times—it was the second curved minimal surface to be known after the catenoid.

Rather than distorting the catenoid, the usual way to obtain a helicoid is to take a straight line L meeting a fixed axis. This axis is the central wire A in the soap film arrangement. The line L is turned with constant motion around this axis A and at the same time is moved uniformly up this axis. Whether we turn in a clockwise or anticlockwise direction will determine whether we produce a right-handed or left-handed helicoid. The end of the line L describes a spiral as turned out on the surface of a cylinder. If we were to take any two points on a cylinder, the shortest distance between these two points would be part of a spiral. Taking both the left-handed and right-handed spirals generates a double helix, so well-known as the model of the DNA molecule.

A variation on this soap-film arrangement is to pass a thread through the wire helix and tie it at certain points along the helix. The film then takes up the minimal surface between the wire helix and the thread by pulling the thread taut as shown in figure 5.9(b).

5.10 CHAPTER APPENDIX; MATHEMATICS OF THE CATENOIDAL SURFACE

Figure 5.10 shows a soap film established between two hoops of radius r_1 and r_2. This enables a more general solution to be obtained than that for hoops of equal diameter. The surface will continue to be symmetric about the x axis and therefore we can consider a cross-section of the surface in the xy plane.

We shall need to integrate with respect to x but the first problem is that the soap film surface is not parallel to x. If we have an element of length along the film ds, this corresponds to an increment of length dx in the x direction, and is given by

$$ds = \left(1 + \left(\frac{dy}{dx}\right)^2\right)^{1/2} dx.$$

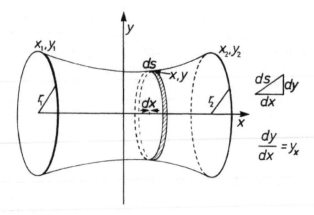

Figure 5.10   Nomenclature for the catenoidal surface.

The notation is simplified if we use y_x to represent dy/dx. The elemental surface area contained between planes perpendicular to the x axis at x and $x + dx$ is

$$dA = 2\pi y \, ds.$$

Hence,

$$dA = 2\pi y (1 + y_x^2)^{1/2} \, dx$$

so that

$$A = \int_{x_1}^{x_2} 2\pi y (1 + y_x^2)^{1/2} \, dx = \int_{x_1}^{x_2} 2\pi f \, dx$$

where f describes a function which depends on y and y_x given by

$$f = y(1 + y_x^2)^{1/2}.$$

Minimization of area requires

$$\frac{\partial f}{\partial x} + \frac{d}{dx}\left(f - y_x \frac{\partial f}{\partial y_x}\right) = 0.$$

This is a well-established mathematical equation called the Euler–Lagrange equation (see for instance Arfken 1966 or Isenberg 1978). Its proof is not given here.

As f does not depend *explicitly* on x, $\partial f/\partial x = 0$, so

$$\frac{d}{dx}\left(f - y_x \frac{\partial f}{\partial y_x}\right) = 0$$

which we integrate to obtain

$$\left(f - y_x \frac{\partial f}{\partial y_x}\right) = c$$

where c is a constant of the integration. Substituting for f gives

$$y(1 + y_x^2)^{1/2} - y_x^2 y(1 + y_x^2)^{-1/2} = c$$

i.e.

$$y(1 + y_x^2)^{-1/2} = c.$$

Rearranging

$$\frac{\mathrm{d}x}{\mathrm{d}y} = c(y^2 - c^2)^{-1/2}$$

$$x = c \cosh^{-1}\left[\frac{y}{c}\right] + c_0$$

where c_0 is a constant. Alternatively,

$$y = c \cosh\left[\frac{x - c_0}{c}\right].$$

Rotating the curve about the x axis gives a catenary of revolution (or catenoid) bounded by the two hoops. c and c_0 are set by the boundary conditions. In the general case, the boundary conditions are established by the radii r_1 and r_2 of the two hoops which are at positions x_1 and x_2 along the x axis. With hoops of equal radius, the origin x can be taken midway between the hoops, giving $y = c$ at $x = 0$. The solution for a film between equi-sized hoops takes on the well-known form

$$y = c \cosh\left[\frac{x}{c}\right]$$

and the value(s) of c is(are) obtained from

$$r = c \cosh\left[\frac{a}{c}\right]$$

where the hoop separation is $2a$.

If the hoops are of unequal radius, an analytical solution is less simple and a numerical solution is preferable.

However, even for equi-sized hoops, the situation is not as simple as the equation might imply. Let us take two hoops with radius unity and separation equal to the radius. As a result of this simplification, the catenoid must pass through points $(-\frac{1}{2}, 1)$ and $(\frac{1}{2}, 1)$ and the equation for the catenoid is

$$1 = c \cosh\left[\frac{1}{2c}\right]$$

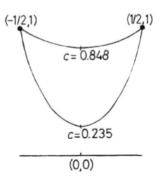

$(-1/2,1)$ $(1/2,1)$

$c = 0.848$

$c = 0.235$

$(0,0)$

Figure 5.11. Two catenaries passing through two fixed points.

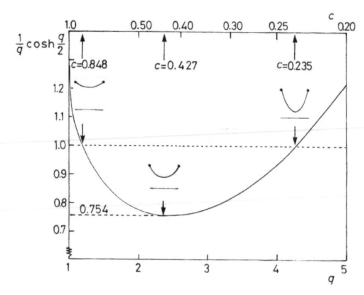

Figure 5.12 $(1/q)\cosh(q/2)$ plotted as a function of q.

which has two solutions: either $c = 0.235$ or $c = 0.848$ (figure 5.11). To see why the mathematics gives two solutions, it is easier to put $q = 1/c$ and plot the function

$$\frac{1}{q}\cosh\left[\frac{q}{2}\right]$$

as a function of q. This is shown in figure 5.12. As well as showing the q values along the x axis, the graph shows the equivalent values of c, which decrease as q increases. As we have seen, the function must equal unity at positions $(-\frac{1}{2}, 1)$ and $(\frac{1}{2}, 1)$. We read off from the graph the two values of q and the two corresponding values of c. That there is a minimum in the curve arises from the opposite trends in the variation of $1/q$ and $\cosh[q/2]$. The two solutions correspond to the fact that two catenaries pass through the points (or hoops or pins). Which catenary corresponds to the film of minimum area? In fact, $c = 0.848$ gives the minimum. It is only if the film is blown through the catenary corresponding to $c = 0.235$ that there is a switch to two flat surfaces (figure 5.13); the second catenary must correspond to unstable equilibrium.

We can see that there will always be a pair of possible catenaries for any hoop or pin positions until we are below the curve in figure 5.12. No stable catenary is then possible. Considering the wedge, suppose we move the pins nearer to the

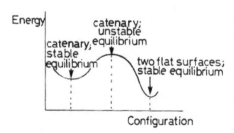

Figure 5.13.   Variation of energy with configuration for the wedge soap-film model.

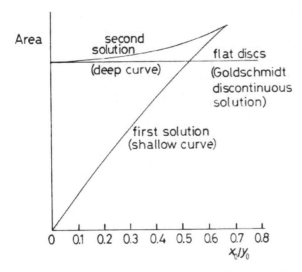

Figure 5.14 Plots of catenary areas as a function of the ratio of ring diameter to ring separation for a hoop (or for the wedge model, the ratio of pin distance from apex to half the separation of the pins).

apex of the wedge to positions $(-\frac{1}{2}, \frac{1}{2})$ and $(\frac{1}{2}, \frac{1}{2})$. We obtain

$$\frac{1}{2} = \frac{1}{q} \cosh\left[\frac{q}{2}\right].$$

for which there is no real solution. There is a limiting ratio of

$$\frac{\text{pin separation}}{\text{pin distance from wedge apex}}$$

corresponding to the minimum of the curve. This occurs for

$$\frac{1}{q} \cosh\left[\frac{q}{2}\right] = 0.754$$

i.e. for a pin position of $y_0 = 0.754$, when the pin separation is unity. This gives

$$\frac{\text{pin separation}}{\text{pin distance from wedge apex}} = \frac{1}{0.754} = \frac{1.326}{1}.$$

In general, the pin position y_0 is given by $y_0 = c \cosh[x_0/c]$ where the pin separation is $2x_0$. If we plot the catenoidal area as a function of the ratio x_0/y_0, we get a variation of the area with x_0/y_0 of the form shown in figure 5.14 (after Arfken 1966).

The first solution, the shallow-curve solution, corresponds to the stable catenary taken up by the film. The second solution corresponds to the unstable equilibrium solution, and the catenary is a much deeper curve. For $x_0 = 0.663y_0$, the area of the complete catenoid is significantly greater than the area of two flat discs. The two-disc solution is often called the Goldschmidt solution. The physics of this situation is that the soap film can no longer maintain the same horizontal force over each vertical section.

For $x_0 < 0.528y_0$, i.e. the separation of the pins is less than 1.056 times the perpendicular distance from the wedge vertex, the catenoidal surface is an absolute minimum.

For

$$0.528y_0 < x_0 < 0.663y_0$$

or

$$1.056y_0 < 2x_0 < 1.326y_0$$

the catenoid is a local minimum, and the film can be easily blown into the two discontinuous discs.

In the case of hoops, we have assumed them to be coaxial. Rather than increase the perpendicular separation, we could move one hoop sideways within its own plane, whilst keeping the other hoop fixed. This movement also gives rise to a situation of instability. The maximum perpendicular separation of the hoops to maintain stability, i.e. $2x_{0m}$, reduces with decreasing overlap of the projected areas of the hoops. The maximum hoop separation reduces to zero as the overlap becomes zero.

6

Soap films within three-dimensional frameworks and minimal surfaces

Having looked at soap films constrained within parallel plates and within a wedge, we now consider equilibrium film patterns established within simple wire frameworks. Some of these frameworks can be constructed to change size. This can lead to unexpected switches between different soap-film patterns. Classification of certain frameworks is undertaken on the basis of deformation of surfaces drawn from the centre of a sphere to great arc configurations on the surface of the sphere. The importance of computers for the graphical representation of minimal surfaces is explored and the derivation of minimal surfaces from mathematical equations is discussed.

6.1 FILMS WITHIN SIMPLE PRISM-SHAPED FRAMEWORKS

Instead of confining films between parallel perspex plates and sets of pins, we can construct wire frameworks which we dip into the soap solution. The film joins up all sides of the framework and has minimum surface area. Of course, it may be that when the framework is dipped only some of the sides are joined, but alternative interesting patterns can result. Although the frameworks are best constructed in a permanent form with stiff wire, it is possible to make them of plastic or from matchsticks. Construction of frameworks is discussed more fully in Appendix I. Even with very simple frameworks, complex and startlingly attractive film patterns can emerge as they are withdrawn from the soap solution.

One of the simplest shapes that we can consider is the *triangular prism*. What pattern emerges depends on the height h of the prism. Provided the height h is greater than approximately $0.5l$, where l is the length of the triangular side of the prism, the

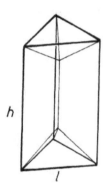

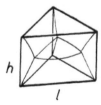

Figure 6.1. Soap-film patterns in the triangular prism.

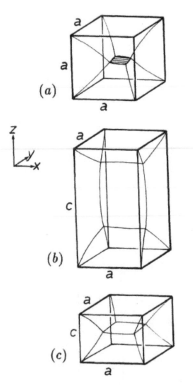

Figure 6.2. Soap-film patterns in the cube and cuboid.

film comprises six triangular components and three trapezia. All nine surfaces are totally flat. The angles between the lines formed from the intersection of these sections are always 109°28′, the so-called Maraldi angle. There are always four lines meeting at a point and the Maraldi angle between any two of these arises from the condition that they meet symmetrically. If the height h is less than approximately $0.4l$, a very different soap-film surface arises. All sections have curved edges, and of the ten different sections making up the total film, four are planar and six are curved in three dimensions. The two alternative film configurations within the triangular prism are shown in figure 6.1. Between approximately $0.5l$ and $0.4l$, either configuration can occur. We consider the cross-section of the soap-film pattern at height $h/2$ for each of these configurations. Then we can see the comparison between the first configuration and the film established using three pins and parallel perspex plates as discussed in Chapter 2 (page 13) and shown in figure 2.1. The second configuration is analogous, although less precisely, to the two-dimensional pattern set up within three pins when a single bubble is enclosed (figure 2.11, page 22). We shall return to the topic of the triangular prism later.

Another simple framework is the *cube*. An equilibrium soap film set up within this framework is analogous to the film in two dimensions between four pins. A flat 'square' section to the film is established (shown as the hatched area in figure 6.2(a)) like the so-called 'bridge' in the 2D case. However, the section is not a perfect square. The edges are curved so that the internal angles are all 120°. Also, whereas in the 2D case the film could be switched between two possible arrangements, here the flat section can be switched between three different planes by blowing. The other twelve sections of the film all show curvature within three dimensions. All sections have to follow the rule that they meet at 120° and that intersections of section edges meet at 109°28′. If the cube can be distorted into a cuboid by increasing or decreasing its height (figures 6.2(b) and (c)) the switching of the central flat section can be induced spontaneously without blowing. This will be discussed further.

The third framework of interest is the *pentagonal prism* (figure 6.3). If the prism is tall, its cross-section at half height is totally analogous to the 2D five-pin case (figure 2.3, page 14). However, if the prism is short, it exhibits a flat pentagonal section at half height; and there are ten other sections within the total soap-film pattern each showing curvature.

In a *hexagonal prism*, the film tries to wrap around the

outside. This is consistent with what happens with the 2D pattern where the film with minimum area goes around the perimeter. However, it is possible to form a film which has a flat hexagonal section at half height that is parallel with the end faces. Twelve other sections of the film extend back from the edges of the hexagon to meet the end edges of the hexagonal prism.

6.2 FILMS WITHIN SIMPLE VARIABLE FRAMEWORKS

The three-dimensional frameworks considered so far have been chosen because it is relatively easy to construct them with variable prism height. Except for the hexagonal prism, all the prisms described allow a switch in configuration to occur by changing the height h. In the case of the *triangular prism*, if we decrease the height h from a large value, this causes the film suddenly to switch configurations when the pair of apexes within the end sections of the film meet.

To allow the film to switch back, the prism height must be increased by a significant amount. Only when the flat triangular section (with curved edges) becomes a small area does the switch back occur. However, this area is not negligible when the switch occurs. Because the edges and hence the surfaces of the film components are curved, a position is reached in closing up the framework when the film spontaneously starts to change shape to form the other configuration. The question of stability of curved surfaces has been discussed already in respect of catenoidal surfaces in Chapter 5.

As was the case with the two-dimensional soap-film models, the configurational changes of the films are analogous to *phase transitions*. On either side of the transition we have different geometrical structures analogous to those occurring in two different crystalline phases. The fact that the changes do not occur at the same value of framework height h, as h decreases and as it increases, is once again an example of the common phenomenon called *hysteresis*.

We can look at the similar situation which occurs if we alter the shape of the *cubic framework* by extending it along one axis. We then have a 'square prism' which is sometimes referred to as a cuboid and sometimes as a tetragonal framework. In crystallography, a crystal is made up of tetragonal cells if the structure can be built up from repeat units of sides a, a and c, where $a \neq c$, and where all three axes are at right angles

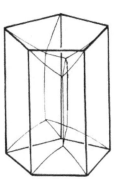

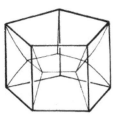

Figure 6.3. Soap-film patterns in the pentagonal prism.

to each other, as in the case of Cartesian co-ordinates. So we make the length of one side of our original cube variable. This length is equivalent to height h but in this case we will call it length c to conform with the notation in crystallography. Altering the length c of the three-dimensional framework is analogous to increasing the separation of the pairs of pins in the four-pin demonstration of a first-order transition given in Chapter 3. Variation of framework shape produces a similar set of film-area versus configuration plots to those shown in figure 3.2 for the 2D case. However, the film areas in this 3D case are difficult to calculate precisely.

To make the comparison, let us start with c significantly smaller than a and with a soap film joining *all* the edges (figure 6.4(a)). The film will consist of triangular and trapezoidal surfaces but with a central section which is approximately square. As we have mentioned already, the edges are curved in order that they meet to produce internal angles of 109°28′. It is the calculation of the curvature of these sides which is not easy to achieve.

In particular, it is the central area we need to watch. If we look approximately side on, this 'square' section appears almost as a single vertical line. The complete film has a total area which is proportional to the total energy of the soap film. The area is the smallest that is possible to join up all the edges of the framework. All film sections meet at 120°, in addition to all edges meeting at 109°28′ consistent with area minimization. The film is stable and cannot be blown into a more stable configuration. If we perturb the film by gently blowing it out of shape, its area will increase, but in due course the film will return to its original position. This is because the film is sitting at the bottom of an energy valley as indicated in figure 6.4(a). Blowing on the film pushes it up the side of the valley and then the film relaxes back down again

Now what happens if we open out the framework? The 'square' section gets smaller and figure 6.4(b) shows the situation when c has been elongated such that it is equal to a. The central 'square' could exist in either of two alternative planes at right angles to the one shown; but it cannot switch unless we cause this to happen by blowing on the film. So we see the film sitting in an energy minimum, with a second equivalent minimum alongside, these minima representing two alternative possibilities. (To include the third minimum here merely complicates the figures.) Elongating the framework further by making $c > a$, we see the central 'square' get even smaller, but remain stable (figure 6.4(c)). It is only when the

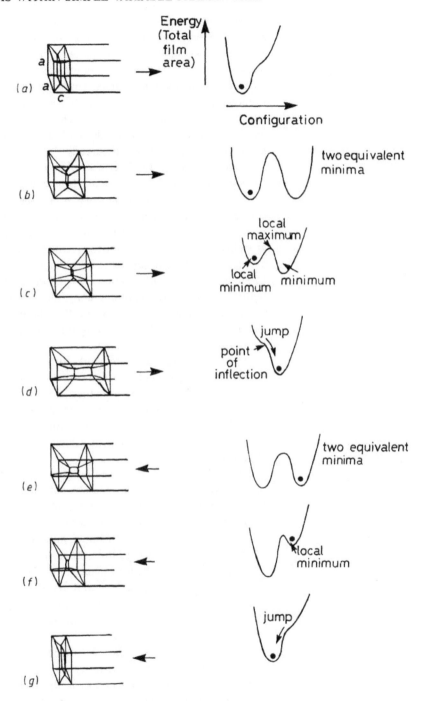

Figure 6.4 Change of film pattern within a variable cuboid. For a description of the stages (a) to (g) see text.

'square' section of film approaches (but does not quite attain) zero area, corresponding to c being approximately $1.2a$, that the energy hillock between the separate minima disappears. Now the film jumps. The central component flips through 90° and takes on a 'rectangular' shape (figure 6.4(d)). Assuming the framework has a cross-section that is precisely square, two alternative configurations of the rectangle are possible, at right angles to each other. The edges of the rectangle meet at the usual 109°28′ angle and are curved.

Next we reduce the length c. As the framework closes up, so does the central 'rectangular' film component. Eventually, $c = a$, such that the framework is a cube. We again see a central film component which is 'square'. At this point the film once more could exist with a central 'square' component at right angles to the present one. Unless we blow on the film, as was suggested earlier, there is no way that the film can jump to the alternative configuration. It is sitting in equilibrium at the bottom of an energy minimum alongside an equivalent one. This is shown in figure 6.4(e). We see that the energy-configuration curve is identical to that shown in figure 6.4(c), but that the film is sitting in the other minimum.

If we close up the framework further such that c becomes smaller than a, the central 'square' section also closes up further. The film could take on a smaller area if it were to jump to the alternative arrangement described, but it cannot do this until this central section closes up almost completely. So once more we see the film sitting in a local minimum-energy valley (figure 6.4(f)).

With the framework closed up sufficiently, the film starts to change shape and slips into the absolute energy minimum. This happens just before the area of the central section becomes quite zero, as the edges of this central section are slightly curved. The criteria of balance of overall forces and of how the total energy of the film changes as the film's shape alters are again important. They determine that the film shape starts to change before closure is complete. The 'square' central section at right angles now opens out. Closing up the framework further produces no other significant change than to increase the size of this central 'square' section.

We have ended up with a set of curves in figure 6.4 which are of similar shape to those in figure 3.2. What we mentioned, but have not taken into account, is the fact that there are two alternative and equivalent arrangements for the soap film when c is large. The 'rectangular' cross-section can exist in two planes at right angles. When the size of the frame is altered, it is random

---

Box 6.1   Twinning

Some crystals, when they are cooled, show phase transitions in which a c direction switches through 90° in a similar manner to that in which the central section of the soap film switches. Hence one ends up with domains in the crystal with the c direction oriented back and forth through 90°. This is an example of a phenomenon called twinning.

A twinned crystal is a composite crystal in which two parts have different orientation, but the operation of a rotation or a mirror reflection brings the orientation of one part into congruence with the other. This can occur if the crystal 'forgets' its patterning during growth. For instance, a crystal built up of A,B,C;A,B,C;A,B,C layers might grow as A,B,C;A,B,C,B,A;C,B,A with a mirror plane at the position of the central C layer (see figure 6.5). Such a twin can be found in face-centred cubic crystals which can be shown to consist of repeated A-, B-, and C-type layers of atoms, where A, B and C characterize the positioning of the atoms.

In the case of the soap film considered in the main text, it can exist with a central rectangle in one of two planes at right angles. These are either a central yz plane or a central xz plane. It is as if a mirror plane exists at 45° to the x and y axes, with this mirror plane including the z axis.

---

which plane is chosen by the soap film. How the framework is held and the effect of gravity can be important. However, a small distortion of the frame can easily give a preferred film orientation. The film can choose very precisely the orientation that gives minimum energy.

The problem of working out the area of the soap film within a cube looks relatively simple but it is not. Even the intersection line between two components of the soap film does not produce a simple curve. The film minimizes its area over the thirteen components. Kathleen Ollerenshaw, a geometrician, calculated the area of the film by an approximation method taking into account the shapes of the various components. She obtained an area of $4.2426a^2$ assuming a cube side of a (Ollerenshaw 1980). This compares with $6a^2$ for a film covering the six

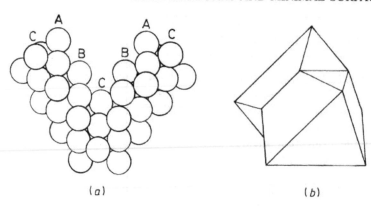

Figure 6.5 Twinned crystals: (a) model; and (b) typical appearance.

faces. These two values produce a ratio of 0.7072:1; quite remarkably $\sqrt{2}/2$. One might argue, as we have done with the two-dimensional patterns, that we need not cover all the faces in order to link all the sides. If only five faces of the cube are filled, the ratio becomes 0.8484:1. The internally connecting film surface remains the one of smaller area. If two opposite faces are unfilled, the ratio is 1.0606:1 and this arrangement would require a smaller area of film. Such a covering does not tend to come about in withdrawing the framework from the soap solution.

The *pentagonal prism* also demonstrates a switching back and forth between its two principle configurations. Starting with a short prism and extending its height, the flat pentagonal section at mid-height becomes smaller and smaller in area. When this section reaches almost zero area, the film switches to the configuration that has two vertical rectangular components (figure 6.3 and frontispiece photographs F2). These two vertical components can take up five alternative positions relative to the framework so that the soap-film pattern now has five-fold degeneracy. Which configuration is taken up depends very critically on any imperfection within the frame.

Thus, the triangular prism, the cuboid and the pentagonal prism are three-dimensional frameworks that can be used for demonstrating features of phase transitions, but not the hexagonal prism because of the tendency of the soap film to wrap around the sides.

6.3 FILM PATTERNS WITHIN THE REGULAR PLATONIC POLYHEDRA

There is a range of other regular frameworks in which interesting and instructive patterns are established. The simplest framework of all, and one not considered so far, is the *tetrahedron* which is a so-called Platonic polyhedron. Within this framework, the film establishes itself as six very precise triangular components, all fulfilling the basic requirements that the angle between adjacent sections is 120° and between lines of intersections is 109°28′ (figure 6.6). We can calculate the area of the film rather easily. Each triangular component has area obtained by multiplying half the size of its base by its height to give

$$(a^2/2)\tan(35°16′) = \left(\frac{0.7071}{2}\right)a^2 = 0.3535a^2$$

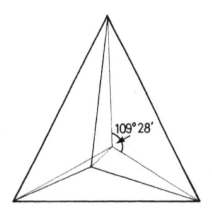

Figure 6.6. Soap-film pattern within a tetrahedron.

where a is the length of each side of the tetrahedron. Hence the total area of the film is $2.121a^2$.

Moving on from the simplest Platonic polyhedron, the next simplest is the cube which has been considered already. The third Platonic solid is the *octahedron*; a soap film can establish a variety of configurations within an octahedrally shaped framework. The most common pattern to be set up consists of twelve triangles and six four-sided platelets each looking rather like a kite. Assuming a side length of a, the triangular areas each have the same area as the triangles set up within the tetrahedron and these twelve give a total area of $4.243a^2$. The area of the six kites is $0.659a^2$ so the total area of the film is $4.902a^2$. The area of a film around the outside of the framework is $4\sqrt{3}a^2 = 6.982a^2$. A number of other film patterns can be set up within the octahedron. However, these exhibit curved intersections between the separate components making calculation of the area difficult. Some of the patterns are highly symmetrical; others are not. Figure 6.7 shows a range of possible patterns of varying stability. The one exhibiting a central non-regular pentagon often appears. Usually more difficult to obtain are those with a square or a hexagonal component at the centre. The octohedral framework therefore provides a particularly good demonstration of different film configurations arising from a series of possible local energy minima. Very often there appears a pattern containing a large rhombus, almost diamond shaped, with slightly curved edges. When this happens, the film is no longer bounded by all the edges of the octahedron.

The fourth Platonic polyhedron is the *dodecahedron* which, as the name implies, has twelve sides. It has thirty edges and twenty vertices. The minimum soap-film surface within the dodecahedral framework is a little different from previous examples. The film effectively breaks away from one side of

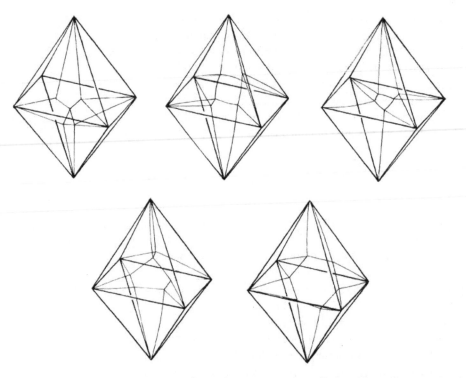

Figure 6.7   Various soap-film configurations connecting all the sides of an octahedron.

the framework to form regular pentagonal faces that are smaller than the pentagonal faces of the dodecahedron. But the film links with all the edges by a series of trapezoidal components.

The last of the five Platonic solids is the *icosahedron* consisting of twenty faces (all triangular), thirty edges and twelve vertices. Now the soap film is quite determined to fill the outer faces, this being a very distinct minimal area into which the film easily slips. The dihedral angle is 138°11′ which is significantly greater than 120°.

6.4   GREAT ARC CONFIGURATIONS

So far we have introduced and discussed a range of minimal surfaces as found to arise within our experimental frameworks. Following on from Plateau, mathematicians have attempted to proceed more systematically in solving Plateau's problem and in obtaining mathematical expressions for minimal surfaces. As has been mentioned, the first curved minimal surface to be recognized was the catenoid and the second was the helicoid.

A different insight into many of the three-dimensional film patterns in this book can be obtained by considering possible film arrangements bounded by great circle arcs as drawn on the surface of a sphere (Almgren and Taylor 1976, Hildebrandt and Tromba 1985). The work originates from Lamarle, a collaborator of Plateau. Great circles are equivalent to lines of longitude and to the equatorial line on the surface of the earth. The approach can be used for a mathematical proof of Plateau's second rule as stated on page 9.

Suppose that we have a system of minimal surfaces and that their lines of intersection are known to be smooth curves. We take any point P within the system of surfaces and consider microscopically smaller and smaller regions on this point. If we have overall area minimization, we still must maintain area minimization however small a region we take from the original. Effectively, our surfaces must in the limit look flat. Taking P as centre, we draw a sphere which will intersect any flat surfaces in the neighbourhood of P. But we cannot intersect more than three as otherwise the surface area can be reduced. We have seen this experimentally within the three-dimensional models. We can have four film components meeting at a point and no more. Here, P is within the surface of one component which accounts for one surface area; so if we only consider the *neighbourhood* of P (and do not therefore extend to regions where the film subdivides at further junctions), it is clear this condition of only three further surfaces follows.

We now consider how many networks of arcs of great circles can be drawn on a sphere such that they always meet at equal angles of 120°. Mathematicians tell us from spherical trigonometry that there are only ten. Finding out how many there are is the same as solving Steiner's problem (section 2.6) on a spherical surface. Next we take cones coming out from the centre of the sphere to intersect the surface along these arcs. The question is whether soap films of minimum surface area can exist congruent to the surfaces of the cones. As the surfaces of the cones meet up at 120° at their edges arising from the constraint on the network of arcs, they do meet the intersection requirements at the surface of the sphere. To outline the cones, and hence the possible film surfaces, we draw straight lines from the centre of the sphere to all points on the arcs. Next we think of possible deformations of these initial film components that might minimize the overall area. What we do not allow is any change of the original great arc configurations on the surface of the sphere.

It turns out that such deformations are possible in all but three cases (figure 6.8). In other words, in three cases the cones are area minimizing, and in the other seven cases deformations must be carried out to produce area minimization.

The first great arc pattern (i) is simply a single great circle. The film is a circular disc and we have already demonstrated the minimization of energy associated with a disc (page 4). The next example (ii) consists of three halves of great circles arranged symmetrically on the sphere. The three flat film surfaces meet at 120° which we know to be correct. What has resulted is a film within a wire framework as shown in figure 6.8 that is often used for demonstrations. In the third minimal example (iii) the great circle arcs are those for a spherical tetrahedron, and here again all the film surfaces meet at 120° and the film edges meet up at 109°28'. We can if we wish straighten the arcs to form the sides of a straight-edged tetrahedron, but we cannot change the film configuration to produce a smaller area. Changing the spherical tetrahedron to a straight-edged tetrahedron does not change the underlying geometry or area minimization.

The remaining seven cases, where deformations are now possible and necessary, are as follows:

(iv) A spherical prism arranged over a base which is a spherical triangle. Taking the cones and carrying out the deformations gives rise to a configuration which is equivalent to that found in a triangular prism of suitable base-length to height ratio.

(v) A spherical hexahedron, consisting of six four-sided polygons. Not surprisingly, this is equivalent to the film surface observed within a cubic framework.

(vi) A spherical prism with pentagonal base. This is equivalent to the film established within a pentagonal prism. We considered this earlier (page 92).

(vii) A spherical dodecahedron. To obtain the equivalent soap-film pattern in a linear framework we use a structure consisting of the edges of the fifth Platonic solid, the dodecahedron (see page 99).

(viii) Four congruent spherical quadrilaterals and four spherical congruent pentagons.

(ix) Two quadrilaterals and eight congruent pentagons.

(x) Three quadrilaterals and six pentagons.

Equivalent linear frameworks can also be used for examples (viii) to (x). The tenth example is different to the others in that the form of the deformation required is rather special and not

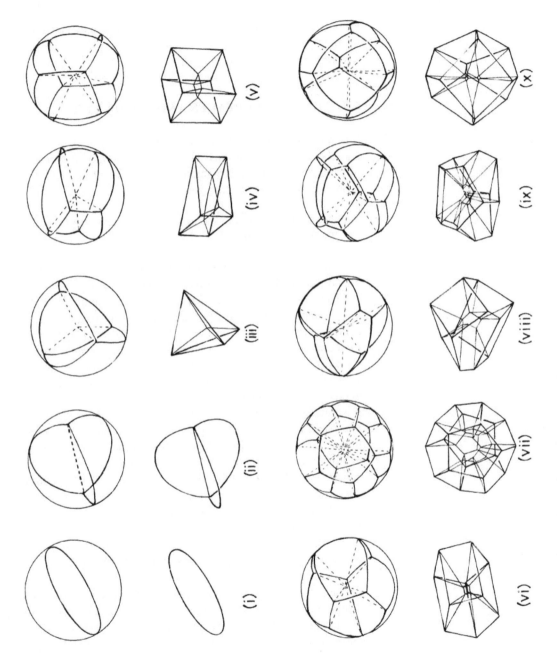

Figure 6.8 Great circle networks and corresponding soap-film configurations.

easily deduced. To save overall area in this case, eleven vertices need to be inserted so as to produce two 'kites' (Taylor 1976).

Although this approach has produced basic surfaces which minimize area, it has failed to produce mathematical solutions to the small curvatures which occur on certain of the faces and intersecting edges.

6.5   MINIMAL SURFACES AND COMPUTERS

Minimal surfaces are of especial interest to mathematicians, not least because it is only possible to find the minimal surfaces to fit certain sets of boundary conditions. The solution of Plateau's problem, the answer to the question

> 'can every closed surface in space be spanned by at least one minimal surface?'

proved elusive for a long time, even though this is a question which intuitively appears to have an obvious answer.

The study of minimal surfaces can be approached from an alternative direction. For all minimal surfaces, the mean curvature at any point on the surface is always equal to zero. Mathematicians are able to set up equations such that this condition is met. What we cannot be sure about is how this surface will look. The arrival of powerful computers and good graphics packages enables the minimal surfaces to be illustrated; see for instance Gray (1993), where the use of *Mathematica* for reproducing curves and surfaces is described. Personal computers have progressed such that they are adequate to illustrate many of the surfaces. The graphics representations can be established within chosen limits and hence boundary conditions judiciously selected. Quite often boundaries can be established within computed minimal surfaces to coincide with the boundaries fixed by frameworks used for the soap-film models. A minimal surface solved mathematically has now become a solution to a problem set up from physics.

Although we are particularly concerned with finite surfaces, mathematicians tend to be interested in minimal surfaces that are infinite in extent and also complete. For a surface to be complete, it must be possible to move around anywhere on that surface in any direction and never come to an edge or to an end. We also require the minimal surfaces to be free from self-intersections, that is the surface must be embedded. (It is perhaps a surprising nomenclature that an embedded surface in mathematics is *free* from intersections.) Most surfaces in nature do not have self-intersections and soap films cannot have them. We can have equations of minimal surfaces and not know whether the surfaces they represent are natural intersection-free surfaces (Callahan *et al* 1988). This is where computer graphics can be particularly powerful. By illustrating different parts of the surfaces, it is possible to see whether self-intersections occur.

Another term often applied to surfaces is *simple*; the topology of the surface is not complicated. Mathematically, a simple surface is one which can be deformed to a sphere to which is added a finite number (which includes zero) of handles or from which is subtracted a finite number (again including zero) of discs. By this definition, a plane is a simple surface, as it is a sphere from which a small disc has been taken. A catenoid is a simple surface because it is a sphere from which two discs are taken. It is now known that one cannot have a complete embedded minimal surface obtained from a sphere with 3, 4 or 5 small discs removed.

Many embedded minimal surfaces are known which are triply periodic (i.e. periodic in the three dimensions of space). Classically, only three non-periodic embedded (i.e. non-intersecting) surfaces are known that are also simple. These are the plane, the helicoid and the catenoid, all surfaces which have been involved in the soap-film discussions. More recently, Callahan *et al* (1988) and Hoffman (1987) have found that a fourth example, due to Costa (1984) and shown in figure 6.9, also meets the stringent conditions. This surface is three-ended, as can be seen from the figure and by comparing the outline of the surface with the outline for a catenoid. Callahan *et al* solved the equations for the surface and then viewed the graphic solutions from a variety of angular aspects. They divided the surface into eight congruent octants and proved that each of these was not self-intersecting. They were able to go on to build a family of examples of higher genus but of similar type.

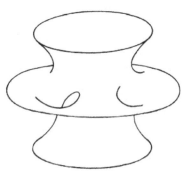

Figure 6.9. The Costa surface.

6.6 EXAMPLES OF MINIMAL SURFACES SUITABLE FOR REPRESENTATION BY COMPUTER GRAPHICS

We have already considered the catenoid and the helicoid as examples of minimal surfaces. A family of minimal surfaces can be described by the parametric equations

$$x = a \sinh \theta \cos \phi - b \cosh \theta \sin \phi$$
$$y = a \sinh \theta \sin \phi + b \cosh \theta \cos \phi$$
$$z = a\phi + b\theta + c_0$$

(see for instance Nitsche (1989, p 70)). We shall not prove the equations here, but they are used in Appendix II to show how to plot both the helicoidal and catenoidal minimal surfaces. For $a = 0$, the equations reduce to those representing a catenoid. For $b = 0$, the equations are for a right helicoid in which the circles of the helix are vertically above each other. See program A.II.12 for the helicoidal surface and program A.II.13 for the catenoidal surface.

There is a form of minimal surface which is particularly complicated in appearance and which was derived from these equations by Scherk in 1835. After a transformation of axes and suitable choice of integration constants, he obtained a minimal surface given by

$$z = \frac{1}{b}\left[\log\left(\frac{\cos y}{\cos x}\right)\right]$$

(see Nitsche (1989, p 71)). The *Mathematica* program for this is given by Dickson (1990) and also Maeder (1991) and is also included as program A.II.14.

We can go on to determine solutions for minimal surfaces represented by a general equation of the form

$$f(x) + g(y) + h(z) = 0.$$

By inserting the condition for mean curvature, plus considerable manipulation including choice of integration constants, it is possible to obtain families of minimal surfaces.

One of the first mathematicians to use such equations to solve Plateau's problem for a non-planar contour was H A Schwarz (1873–1916). He worked on minimal surfaces at the same time as Plateau. Figures 6.10(a) and (b) show surfaces obtained from similar equations, but with the establishment of different boundary conditions. The first surface is one bounded by six (out of the total of twelve) edges of a cube. These boundary edges are the six originating from two opposite corners of a cube. This particular minimal surface can be divided into six congruent areas.

For the example usually referred to as the Schwarz minimal surface, Schwarz used the tetrahedron framework (see page 99), but he was specifically interested in a film which did *not* join all

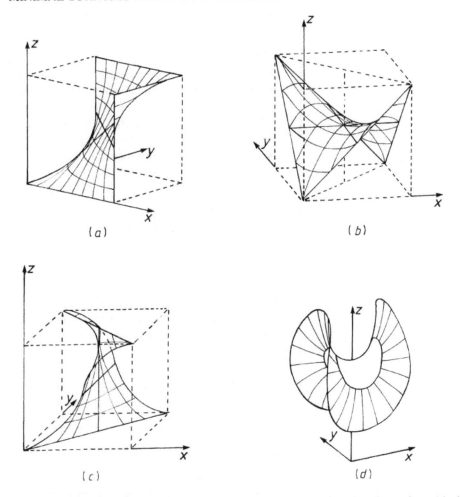

Figure 6.10   Some minimal surfaces: (a) curved minimal surface joining six edges of a cubic framework; (b) Schwarz surface within a tetrahedron and its relationship to the cube; (c) Gergonne's surface; (d) part of Enneper's minimal surface.

the edges. He was investigating a film which joined four of the six edges (figure 6.10(b)) and produced a semi-permanent model using a thin wire framework and a skin of gelatine. This model illustrates especially well a surface exhibiting a saddle point (cf. the saddle point shown in figure 1.6(c) on page 10). It might perhaps be truer to say that, within a minimal surface, which Schwarz found suitable for study and solution, it turns out that we can insert straight line boundaries. These then correspond to the particular framework boundaries in the real-life example of a soap film within a tetrahedron.

Just as we have seen how two-dimensional soap-film patterns

Box 6.2 Schwarz minimal surface

There are many mathematical representations of this surface. All are equivalent but perhaps the simplest to write out is that which uses complex number notation with $i = \sqrt{-1}$ and 'Re' meaning that we should take the real part of what follows after the expression. The surface can then be written in the form

$$x = \mathrm{Re} \int^{\omega} (1 - \omega^2) R(\omega)\, d\omega$$

$$y = \mathrm{Re} \int^{\omega} i(1 + \omega^2) R(\omega)\, d\omega$$

$$z = \mathrm{Re} \int^{\omega} 2\omega R(\omega)\, d\omega$$

where $\omega = u + iv$ and $R(\omega) = [1 - 14\omega^4 + \omega^8]^{-1/2}$. There is a complication in that $R(\omega)$ has singularities (there will not be physical solutions) at $\omega = \pm(\sqrt{3} \pm 1)/\sqrt{2}$ and $\pm i(\sqrt{3} \pm 1)/\sqrt{2}$.

    A further minimal surface of similar type is called **Gergonne's surface** and is shown in figure 6.10(c). This can also be broken down into four congruent parts.

Box 6.3 Catalan's surface

There is a minimal surface related to the catenoidal surface that has equations (see Nitsche (1989, p 141)) of the form

$$x = \alpha - \sin\alpha \cosh\beta$$
$$y = 1 - \cos\alpha \cosh\beta$$
$$z = 4\sin(\alpha/2)\sinh(\beta/2).$$

The surface contains a cycloid (with parametric equations $x_0 = \alpha - \sin\alpha$, $y_0 = 1 - \cos\alpha$, and $z_0 = 0$) as a geodesic (i.e. a curved line lying within the minimal surface). Such a surface is called Catalan's surface and the program to plot this surface is given as A.II.15.

Box 6.4   Enneper's surface

A more general minimal surface than Catalan's surface is Enneper's surface.   It can be obtained from a generalization of the equations for Catalan's surface, but alternatively the surface can be defined mathematically by starting from the equations for two parabolae with focal points lying in orthogonal planes, as follows:

$$x^2 = \tfrac{8}{3}z + \tfrac{8}{9} \qquad y = 0$$

and

$$y^2 = \tfrac{8}{3}z + \tfrac{8}{9} \qquad x = 0.$$

The vertices and focal points of the parabolae alternate in their coincidences.

From these equations for the parabolae two sets of parametric equations are obtained for x, y and z, one set involving variable u and the other variable v, which become the parameters of the minimal surface.   By mathematical manipulation, equations for the surface can be given in polar coordinates in the parametric form

$$x = r\theta - \tfrac{1}{3}r^3 \cos 3\theta$$
$$y = -r \sin \theta - r^3 \sin 3\theta$$
$$z = r^2 \cos 2\theta.$$

The appearance of what we see as the Enneper surface depends very much on what portion of the curve we draw. It can be shown that the piece of Enneper's surface equivalent to $r < \sqrt{3}$ has no self-intersections. A part of the surface is shown in figure 6.10(d). Program A.II.16 allows a variety of parts of the surface to be drawn out using *Mathematica*.

---

can be broken down in order to simplify calculation of the overall length, so here we can see immediately that the minimal surface can be broken down into four congruent parts bounded by straight line segments. Mathematical calculation of the area, originally discussed by Schwarz (1890), is considered by Nitsche (1974, 1989) and others.

Schwarz discovered two very important reflection principles

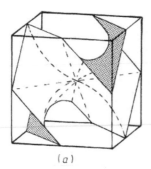

(a)

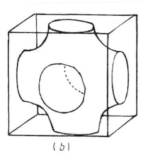

(b)

Figure 6.11. (a) Schwarzian periodic minimal surface (the Schwarz F surface) produced from repeat units illustrated in figure 6.10(b). (b) Schwarz's P surface.

for minimal surfaces. The first is that if part of the minimal surface contains a straight line, then its reflection across this line is also a minimal surface; the union of these two surfaces then also forms a smooth minimal surface. Similarly, if a minimal surface meets a plane at right angles, then its mirror image at that plane is also a minimal surface, and once again the two surfaces can be joined to form a single smooth minimal surface.

We can see how the surface can be repeated, in this case as many times as we wish, to produce a more complicated minimal surface. After each reflection and incorporation of the mirror image, we repeat at the next straight line. This can be expected to have general application, providing the reflection is over a full unit of the framework; but in some examples the resulting surface leads to self-intersections. In other special examples, Schwarzian chains bound minimal surfaces which extend forever without intersection. The resulting surface is called a periodic minimal surface. The Schwarz surface within the tetrahedron can be used to construct such a periodic minimal surface and part of this surface is shown in figure 6.11(a). It is called the Schwarz F surface. It is easy to see how further and further sections can be added and also how a whole range of Schwarz minimal surfaces are possible based on the tetrahedral framework. Mackay (1985a,b) has discussed how such surfaces can be related to crystallographic space groups and to crystal structures. Glycerol mono-oleate (GMO) with water has been shown to take on the shape of the Schwarz F surface (Longley and McIntosh 1983).

The so-called Schwarz P surface (figure 6.11(b)) consists of six tubes connecting each closed domain (a cubic box) to the next. The consequence of this is that the surface divides all of space into two congruent regions. This P surface can be used to separate the two primitive lattices on which the caesium ions (Cs^+) and chlorine ions (Cl^-) exist in caesium chloride (CsCl) and the P surface closely resembles the zero electrostatic potential in caesium chloride existing between oppositely charged ions.

These minimal surfaces are similar to Fermi surfaces for electrons in metals. We cannot go into the details of the origin of Fermi surfaces here. It is sufficient to say that Fermi surfaces are drawn in 'momentum space' but that they have a relationship with the regular crystalline lattice because of the interaction between the periodic potential of the lattice and the wavefunctions for the electrons. Here also, the surfaces represent uniform energy surfaces around which the electrons can move.

Figure 6.12 shows the Fermi surface for copper or gold. It can be seen that this surface closely resembles the Schwarz P surface. The size of the connecting tubes can vary for different materials and the Fermi surface can take on alternative shapes. Outlined in the figure is the so-called dog's bone orbit around which an electron can move when the gold or copper is placed in a magnetic field.

In this chapter we have looked at minimal soap-film surfaces in three-dimensional frameworks and at the computation of minimal surfaces from mathematical equations. We have indicated how these two approaches for developing minimal surfaces can be brought towards convergence. Minimal surfaces have applications for many areas of science. Experimental models to produce soap-film minimal surfaces can have considerable application for cases where theoretical derivations are difficult or impossible at present.

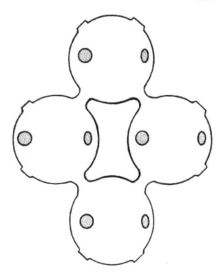

Figure 6.12. Fermi surface for copper or gold.

7

Fermat's principle and refraction

Comparison is made between Fermat's principle for light and the minimization of area by soap films. The variation of height of a film constrained between perspex plates is shown equivalent to the variation of refractive indices of media through which light may pass. The analogy is used to model the refraction of light as it travels through a prism and lenses, and modelling is extended to seismic waves through the earth. The curvature of space and the bending of the path of light close to a black hole are also discussed.

7.1 FERMAT'S PRINCIPLE

In its most commonly quoted form, Fermat's principle says that out of all the paths that light might take to get from one point to another, the path that requires the shortest time is the one that is taken. Although it was found necessary to extend Fermat's principle to the more general statement that the light paths represent stationary values rather than minimum values, the analogy with soap films made here will use Fermat's principle in its original form.

The time taken by light to travel a particular path will be the distance of that path divided by the velocity of light. In a medium the velocity of light is lower than in free space or in air. In vacuum the velocity of light is c and it is very little different to that in air. In a medium of refractive index n the velocity is reduced to c/n. Fermat's principle becomes of particular interest when the light travels through a number of media of differing refractive indices. The light then tries to adjust its path through the media to allow for the differing velocities. If the light passes through media of refractive indices n_1, n_2, n_3, and so on, and the path lengths through each of these media are x_1, x_2, x_3, etc, then Fermat's principle is equivalent to stating that the summation

$$(1/c)(n_1 x_1 + n_2 x_2 + n_3 x_3 + \cdots) \qquad \text{(for light)}$$

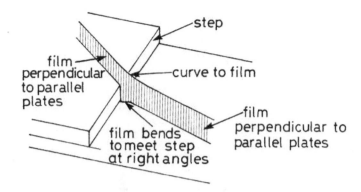

Figure 7.1 Soap-film surface at a step.

is minimized over the distance between the starting and finishing points on the path of the ray of light.

If a soap film is established between two pins set between parallel perspex plates, the film will take up the shortest distance; i.e. a straight line. This minimizes the area of the film and hence the energy of the film. The path is the same as the straight path taken up by light passing between these points. However, we can make the spacing between the plates stepped. We can have separations between the parallel plates of s_1, s_2, s_3, etc with the lengths of soap film between the plates at these respective separations of x_1, x_2, x_3, etc. The film minimizes its area and to do this it minimizes the summation

$$(s_1 x_1 + s_2 x_2 + s_3 x_3 + \cdots) \qquad \text{(for soap films).}$$

Perhaps we should note that the soap film in the region of a step is only approximately in the form of a 'tape' (figure 7.1). Although it consists of a surface that meets the plates and step at right angles, the curvature in the region of each step is non-zero in a direction perpendicular to the horizontal plates. There is a curvature which is shown in exaggerated form in figure 7.1. The curvature will always satisfy the equation

$$\frac{1}{R_1} + \frac{1}{R_2} = 0$$

where R_1 and R_2 are the principal radii of curvatures of the film at any point. This means that minimizing area is not quite a perfect analogy for minimization of optical path. But it is a good analogy.

Whereas we have the refractive index contained as a term within the minimization condition for Fermat's principle,

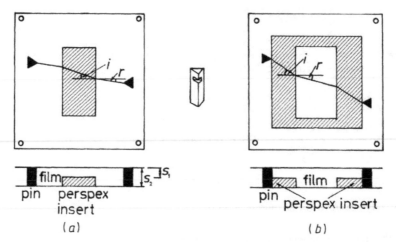

Figure 7.2 Soap-film analogue of refraction using inserts equivalent to (a) refractive index $n < 1$; and (b) refractive index $n > 1$.

we have plate separation as the equivalent term within the minimization condition for the soap-film model. As a consequence, the passage of light through a series of media possessing different refractive indices can be modelled by soap films between plates with a series of different separations. To represent media with larger refractive indices we need increased plate separation.

It is of particular interest to model various examples which demonstrate the refraction of light. If light passes from one medium to another at right angles to the interface, there will be no refraction and no deviation of light from the straight line. Similarly, if the stepped regions crossed by the soap film all have their steps set at right angles to the film, again there will be no deviation. So we model a situation where the light or the film meet the interface at an angle other than 90° to demonstrate refraction.

7.2 MODELLING SIMPLE REFRACTION

The most versatile method of demonstrating the effect of simple refraction is to set up parallel plates separated by spacers and to use movable pins (see Appendix I for more detail) as starting and finishing points for the soap films. Movable perspex blocks between the outer plates are used to give regions of reduced plate separation. Inserting a solid perspex block within the space is equivalent to reducing the refractive index and so is equivalent to cutting out a cavity inside of glass (for example). Hence, the refractive index is given by the ratio of the angle of incidence i to the angle of refraction r as one goes *out from the insert*.

If the spacing between the insert and the upper plate is s_1 and between the upper and lower plates is s_2, (figure 7.2(a)), then

$$\frac{\sin i}{\sin r} = \frac{s_2}{s_1}$$

corresponding to Snell's law for the refraction of light at an interface of

$$\frac{\sin i}{\sin r} = \frac{n_2}{n_1}$$

where n_2/n_1 is the ratio of the refractive indices for the two media. By altering the angle at which the film meets the block, we can see that the ratio for $\sin i / \sin r$ is maintained up to the critical value for r, when the value of $i = 90°$.

To model the more usual situation of light passing from air into a glass block and back out into air again, it is necessary to use a perspex insert with the shape of the block *cut out* (figure 7.2(b)). Note that to represent a refractive index of $n = 1.5$ we require an insert which is one third the separation of the main pair of parallel plates and for $n = 1.33$ we need inserts of thickness one quarter the separation of the parallel plates.

To set up the films, the movable pins are placed in suitable positions to represent the start and end points of a ray of light. The parallel perspex plates are dipped into soap solution and removed. This procedure should establish a single soap film between the pins. The requisite perspex block is then slid into place, having been first wetted. The film should take up the correct equilibrium path. There is a tendency for the film to stick at the corners of the blocks or at the corners of the cut-out sections, in which case the film needs to be blown past these points. Any unwanted films, such as those joining spacers between the plates, can be removed by poking them with a wet drinking straw.

7.3 PRISMS AND LENSES

We can demonstrate refraction in a prism made of a less dense medium set in a more dense medium of higher refractive index by using a triangular perspex insert (figure 7.4(a)). It is more realistic to model a prism made of dense material such as glass surrounded by air (or vacuo). We do this by using a perspex

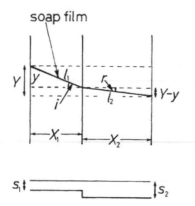

soap film

Figure 7.3. Illustration of Snell's law.

Box 7.1 Snell's Law

We can check the equations mathematically. We refer to figure 7.3 where a film passes between perspex plates of separation s_1 and then between perspex plates of separation s_2. The film meets the interface between the two plate separations at an angle unequal to 90° and changes direction at this point. The film extends for a distance l_1 between the plates of separation s_1 and extends for a distance l_2 between the plates of separation s_2. The angle of incidence is i, such that tan i equals y/X_1 and the angle of refraction is r, such that tan r equals $(Y-y)/X_2$. We need to minimize the total area A equal to

$$l_1 s_1 + l_2 s_2 = \sqrt{(X_1^2 + y^2)s_1} + \sqrt{(X_2^2 + (Y-y)^2)s_2}.$$

The only variable in the above expression is y so we obtain

$$\frac{dA}{dy} = \frac{\frac{1}{2}2y s_1}{\sqrt{X_1^2 + y^2}} - \frac{\frac{1}{2}2(Y-y)s_2}{\sqrt{X_2^2 + (Y-y)^2}} = 0$$

$$\frac{y s_1}{l_1} - \frac{(Y-y)s_2}{l_2} = 0$$

$$s_1 \sin i - s_2 \sin r = 0$$

$$\frac{\sin i}{\sin r} = \frac{s_2}{s_1}.$$

For light following the same path as the soap film above we need to show the minimization of $(t_1 + t_2)$ where t_1 is the time to pass through medium 1 and t_2 is the time to pass through medium 2. This means minimization of

$$\frac{l_1}{v_1} + \frac{l_2}{v_2}.$$

So we need to replace s_1 for the soap-film model by $1/v_1$ and to replace s_2 with $1/v_2$:

$$\frac{\sin i}{\sin r} = \frac{v_1}{v_2}.$$

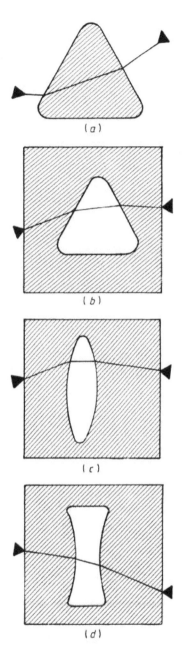

Box 7.1 *Continued*

However, the refractive index n_1 for medium 1 is c/v_1 and the refractive index n_2 for medium 2 is c/v_2, where c is the velocity of light in a vacuum. This gives

$$\frac{\sin i}{\sin r} = \frac{n_2}{n_1}.$$

---

insert with a triangular hole cut from it (figure 7.4(b)) to represent the prism.

Similarly, we can see the effect of a film crossing a curved interface, and particularly the effect of a film passing through cut-outs shaped as lenses. For the case of a converging lens (figure 7.4(c)), it is important to make the cut-out section sufficiently large. The curved surfaces must be of sufficiently large radius of curvature, or the thickness of the perspex insert sufficiently small, for the film to pass through the cut-out section. Otherwise, the film will minimize its area between the pins by passing outside the lens outline. This situation would correspond to constructing a lens of such short focal length that it converges the light close to the exit point. This would be equivalent to converging the light within the region of the lens where one is unable to place a pin. Radii of 10 cm for the curved surfaces are suggested if a plate separation of 1 cm is used, together with perspex inserts of thickness 3 mm or 4 mm. (For further information see the calculation in box 7.3.)

For the analogy with a diverging lens, it is again essential for the soap film to be able to minimize its area when passing through the lens outline (figure 7.4(d)), but this proves to be a smaller constraint than for the converging lens. In each example, the pins fix the starting and finishing positions of the soap film, and the film moves its position within the lens to minimize its area and hence demonstrate the optical path.

7.4 MIRAGES

A medium of uniformly changing density and hence refractive index can be modelled by a soap film within a wedge. The fact that a film curves into the narrower part of the wedge in order to minimize its area has been discussed in detail in Chapter 5.

Figure 7.4. Refraction in: (a) prism with $n < 1$; (b) prism with $n > 1$; (c) convex (converging) lens; and (d) concave (diverging) lens.

Box 7.2   Analogue with light passing
symmetrically through a prism

Let us consider light passing through a prism of apex
angle magnitude A. For simplicity we will consider the
symmetrical case in which the angle of entry at one face
of the prism equals the angle of exit at the other face (see
figure 7.5). We assume the total angle of deviation of the
light has magnitude D. Thus a wavefront XY deviates
through angle $D/2$ at the first surface and a further angle
$D/2$ at the second surface to produce wavefront $X'Y'$.
We can see from the figure that the angle of incidence is
equal to $(A + D)/2$ and the angle of refraction is $A/2$.
Hence,

$$\sin\left(\frac{A + D}{2}\right) = n \sin \frac{A}{2}.$$

For a refractive index of 1.50 and an apex angle of 60°
(i.e. using a prism whose cross-section is an equilateral
triangle), the angle of deviation is 37°. This is the scaling
shown in the figure and is one which is easy to replicate
with the soap-film analogue. Using a 90°, 45°, 45° prism
constructed for $n = 1.50$, the deviation is 25° for an apex
angle of 45°. The 90° apex angle cannot be used as the
angle of incidence would need to be greater than 90°.

Setting up the analogue for a prism containing a liquid
such as water with $n = 1.33$, we obtain an angle of
deviation of 33° for an apex angle of 60° and a deviation
of 16° for an apex angle of 45°.

The discussion there included details of how the film curved in
two directions. The analysis illustrates clearly the fact that the
soap film does not bend as a 'tape'. It has a curvature across
the tape rather analogous to the bend present on many extending
steel rules. So the analogy is perhaps no better than for stepped
spacings where we have seen already that there is distortion of
the film at the step edges. Nevertheless, if the angle of the wedge
is small, and in real life the variation of refractive index will be
very very small, then the analogy is acceptable.

The curved path of the rays of light producing a mirage such
as in the desert arises from the increasing refractive index as
the light moves upwards through the atmosphere starting from

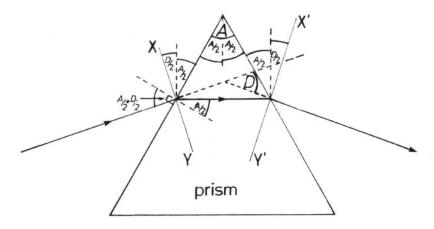

Figure 7.5 Symmetrical refraction of light in a glass prism.

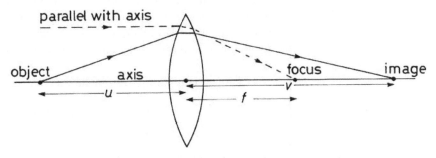

Figure 7.6 Notation for the simple lens formula.

ground level. The sun heats the air just above the sand or the road. This heating of the air produces a lowering of air density and consequently a lowering of refractive index close to ground level. The eye sees the mirage of an oasis in the desert, or the mirage-type effect on a hot road, by looking along the ground at a grazing angle. To establish the equivalent geometry with the soap-film model, the wedge must be stood on its apex, such that the wider section of the wedge corresponds to the larger refractive index at the higher level in the atmosphere (see figure 7.7).

The more practical way to stand the wedge is with its apex uppermost. This corresponds to less dense air uppermost and produces the less common and so-called 'superior' mirage. Whereas the usual 'inferior' mirage is observed over hot land, particularly over deserts, the 'superior' image can sometimes be observed over a cold sea or a frozen surface, but only rarely. Distortions of the appearance of the sun and moon when observed

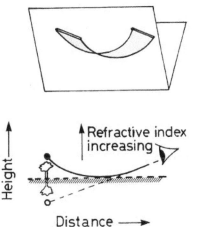

Figure 7.7. The mirage.

Box 7.3   Analogue with light passing through convex and concave simple lenses

---

The standard formula for a lens (see any optics book or basic physics text containing optics) is $1/f = 1/u + 1/v$ where u is the distance between the object and the centre of the lens, v is the distance between the image and the centre of the lens and f is the focal length of the lens (figure 7.6). (If we shine parallel light onto a convex lens, the focal length is the distance between the centre of the lens and the point on the far side where the rays are brought to a focal point.) There is a convention for u and v of positive sign for real objects and images and negative sign for virtual objects and images. The soap film analogy is restricted to real objects and images. For convex lenses, positive values of f apply, whereas for concave lenses f will be negative. However, it is convenient to express the focal length in terms of the radii of curvature of the lens surfaces.

Another standard optical equation is

$$\frac{1}{f} = (n - 1)\left(\frac{1}{R_1} + \frac{1}{R_2}\right)$$

where R_1 and R_2 are the radii of curvature of the two lens surfaces. The equation is sometimes called the *full lens formula* or *the lensmaker's formula*. We use positive values for the R values for the convex lens and negative values for those of the concave lens.

So on the basis of scaling for soap-film lenses with cut-out height (corresponding to lens diameter) of approximately 10 cm and radii for the two surfaces of 8 cm, we can deduce from these figures a focal length of 8 cm for both the convex and concave lenses if we have refractive index n of 1.5, and we deduce a focal length of 12 cm for a refractive index of 1.33. If planar-convex or planar-concave lenses are used, the value of R_1 is infinity and focal lengths increase to 16 cm and 24 cm for the $n = 1.5$ and $n = 1.33$ cases respectively; hence it is preferable in this case to reduce the radius of curvature of the non-planar side.

at rising and setting can also arise from the bending of light rays due to the variation of refractive index of the atmosphere. For further information on mirages see Greenler (1980).

7.5 P-WAVES THROUGH THE EARTH

The modelling of the passage of waves through media is not restricted to light waves, or even to electromagnetic waves. We can model the passage of sound waves and this can include seismic waves which pass through the earth as a result of an earthquake. Seismic waves can be either P-type which are longitudinal waves (we might call them 'push–pull' waves) or S-waves which are transverse or shear waves. The first type of wave is analogous to sending a wave of compressions and rarefactions down a slinky. The second type is equivalent to sideways motions passing down the length of the slinky.

In considering the refraction of either sound waves or seismic waves, we need to consider the change of angular direction at an interface in terms of the ratio of velocities in the two media either side of the interface rather than in terms of the ratio of refractive indices. We use Snell's law in the form $(\sin i / \sin r)$ equals (v_1/v_2) already discussed. The velocities of the seismic waves will depend on their type and the nature of the media through which they pass (see later).

7.5.1 The earth's structure

First we consider the structure of the earth (see, for example, Open University 1987). The outer region of the earth is called the crust and this region is relatively thin. Since it is only 90 km thick under continental regions and considerably less under the oceans, representation in any model requires considerable exaggeration of its thickness. Next comes the mantle of the earth separated from the crust by the so-called Mohorovičić discontinuity. This discontinuity was discovered when it was found that waves could travel by two different paths from an earthquake epicentre (the central region of the earthquake) to a recording centre not too far away. One path goes directly between the epicentre and the recording station via the crust, but the other path is via the mantle as shown in figure 7.8(a). This path is possible because of the distinct difference in the velocity of the waves when passing through the mantle and

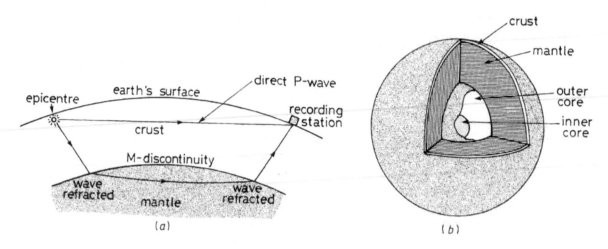

Figure 7.8 (a) The Mohorovičić discontinuity showing how a crustal P-wave and a refracted P-wave can be recorded at a distance of several hundred kilometres from the epicentre of the earthquake. (b) Structure of the interior of the earth.

through the crust; this produces a strong refraction of the waves at the Mohorovičić discontinuity.

The change in density arises from a change from granite (essentially where there are continents) or basalt (under the oceans) to peridotite which occurs deeper down. Granite is a feldspar and peridotite is composed of a range of minerals, chiefly pyroxene and olivine, in varying quantities. All these minerals have structures based on SiO_4 (i.e. silica plus oxygen) tetrahedra, but the precise crystal structures vary, as do the metals incorporated within the tetrahedra. Taking into account this variation of structure and mass, it is not surprising that the seismic waves are transmitted through the various media with differing velocities.

The mantle itself extends deep into the earth; it extends for more than a third of the earth's radius inwards. Both the density of the mantle and the velocity of the waves increase with greater depth (see pages 125–6). Strictly, the mantle should be divided into the upper mantle and the lower mantle with lower and higher average densities respectively; there are in addition regions of different rock densities within these subdivisions. However, these variations do not affect the overall arguments regarding the passage of waves through the mantle. The variation of density arises from gradually changing pressure within the earth and also from sudden chemical changes and phase changes.

Below the mantle comes the outer core (figure 7.8(b)) which

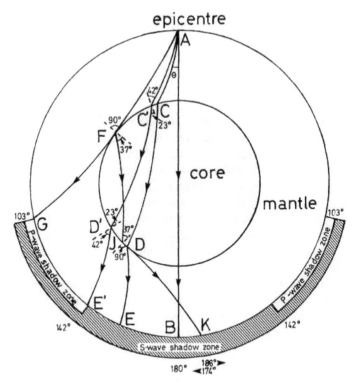

Figure 7.9 Passage of S- and P-waves through the interior of the earth, ignoring the inner core and showing the shadow zones.

is a mixture of iron and sulphur (probably) and is *liquid*. This last fact is very important as, although longitudinal P-waves can pass through this region, shear S-waves cannot. It is for this reason that we will confine arguments for the soap-film model to P-waves. In the meantime, in discussing the waves within the earth's interior, we shall continue to consider both types of wave. Finally, the inner core is solid. It consists of a mixture of iron and nickel.

7.5.2 Passage of P- and S-waves through the earth

Let us see what happens to the P- and S-type waves as they are set off from an epicentre at point A close to the surface of the earth (figure 7.9). We shall consider what happens at the mantle/outer core boundary. We will omit the crust, which has negligible effect, and the inner core which we will insert later. A P-wave can travel radially downwards and be detected undeviated on the other side of the earth at B by meeting each

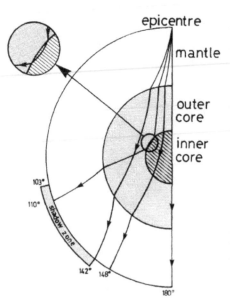

Figure 7.10. Effect of the inner core on the passage of S- and P-waves through the interior of the earth. (The inset shows an enlargement of the path for critically refracted waves at the outer core/inner core interface.)

interface at an angle of incidence of 0°. Similarly, waves making an angle θ to the vertical will also arrive somewhere on the other side, but there will be some bending of the path to allow for the variation in density of the rocks through the different regions of the earth, and also refraction at the mantle/outer core boundary as the wave enters and leaves the outer core. A typical path is ACDE.

However, what happens to other waves needs careful calculation. As θ increases, we reach a situation AC'D'E' where the P-waves emerge at an angle of 142° as measured around the surface of the earth from the epicentre at 0°. For higher magnitudes of θ, the consequent geometry, arising from the spherical nature of the mantle and core of the earth plus refraction at the interface, results in the seismic waves being bent back further into the earth. This leads to the waves emerging between E' and B and even beyond B. Eventually, there comes a point where the angle at which the wave meets the mantle is equal to the critical angle (beyond which the waves are internally reflected back), and this is the most extreme path possible through the outer core (path AFJK). This path emerges at 186° as measured in an anticlockwise direction, or 174° as measured clockwise. Increasing θ further, we arrive at a possible path AFG such that the waves pass along a curve through the mantle, but the waves never enter the core. Between E' and G we have a shadow zone for the P-waves. In practice, this lies between angles 103° and 142° as measured angularly in an anticlockwise direction around the surface of the earth from the epicentre. Because of the molten nature of the outer core, the S-waves are shadowed from 103° right around to 180° and onto the 103° point, as measured in the clockwise direction around the earth.

This has modelled the paths of the waves through the earth reasonably well. However, we have not considered the possible effect of the inner core. The resultant modification to the paths of the P-waves is shown in figure 7.10. There is little effect on the emergence of P-waves except that there is a small region at approximately 110° where a P-wave can be deviated within the inner core and emerge within the shadow zone. The seismologist's task is made even more difficult in that the P-waves can weakly generate S-waves in the inner core which can be regenerated as P-waves at the interface to the molten outer core.

7.5.3 The velocities of the waves

We have indicated that the refraction of the seismic waves is

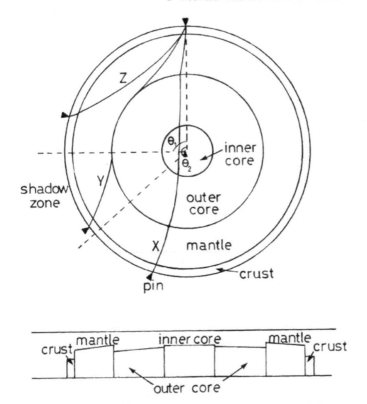

Figure 7.11 Soap-film analogue for the passage of P-waves through the earth's interior.

best described in terms of the ratios of velocities within adjacent media. We have not mentioned how the velocities vary according to the properties of the minerals making up the different layers. For P-waves, their velocity is given by

$$\sqrt{\frac{\text{axial elastic modulus of medium}}{\text{density of medium}}}.$$

How the velocity varies with depth depends on whether the axial elastic modulus or the density varies most rapidly. An increase in density (as one might expect as one goes deeper into the earth) would produce a decrease in velocity. This is not what happens. So the axial elastic modulus must increase more rapidly than the density decreases, in order to produce the increase in velocity. As the structures of the minerals become more rigid at greater depth within the earth, the waves travel considerably faster. For S-waves, the axial elastic modulus used in the expression for P-wave velocity is replaced by the shear elastic modulus.

Table 7.1 Data used for modelling the interior of the earth.

Region	Distance from centre of earth (km)	Velocity (km s^{-1})	Height of perspex (mm)
Crust	6370	6.7	8
Mantle	6370 / 3470	8.1 / 13.6	11.5 / 16
Outer core	3470 / 1215	8.1 / 10.3	11.5 / 14
Inner core	1215 / 0	11.2 / 11.2	15 / 15

7.5.4 Modelling the paths of P-waves with soap films

Obviously, we cannot achieve all these features in a soap-film model but we can show the main ideas. We construct an insert consisting of a series of concentric circular sections shaped to provide different and changing spacings. Figure 7.11 shows such an insert. The cross-section of the insert has been constructed such as to leave appropriate air spacings when the insert is positioned between appropriate parallel plates. The greatest limitation is that it is not possible to model internal reflection which is an important feature for an overall picture of the passage of waves through the earth's interior.

Table 7.1 shows the main regions of the earth, their distances from the earth's centre and the velocities of the P-waves (the longitudinal waves) in these regions. Included in the table are suggested dimensions for the perspex model for a plate separation of 22 mm. The mantle and outer core have been assumed to have a uniformly changing density and have been constructed with a uniform taper. The 22 mm plate separation is somewhat arbitrary. Changing this separation would change the heights through the cross-section of the insert. However, a wider plate separation for this application is desirable than for the parallel plates used with the previous models. This is to allow suitable scaling.

A number of alternative paths for the soap film through the model are shown in figure 7.11. Path X shows the film passing through the inner core to emerge at a point on the far side of the disc. If the bottom pin is then moved in a clockwise direction, the film moves to the outside of the inner core and then sweeps across the outer core to take up path Y. This path is not an actual path for P-waves through the earth's interior. In practice, the waves are reflected at the boundary between the mantle and outer core and so path Y in the model is emerging into what would be the shadow zone on the earth (figures 7.9 and 7.10). Path Y does correspond to a wave path which is a stationary value according to Fermat's principle, but the amount of energy transmitted along the path will be negligible. So the amount of energy transmitted along this path within the earth will be undetectable.

On moving the pin further round clockwise, a position is reached where the film passes through the mantle alone without touching the interface between the mantle and the outer core. Paths of this type (for instance path Z) emerge beyond the shadow zone. In the case of P-waves passing through the earth, we saw that the shadow zone lies between angles $\theta_1 = 103°$ and $\theta_2 = 143°$. For the model as described, it is found experimentally that these angles are somewhat smaller. Such a model cannot accurately reconstruct the density variations within the earth's interior and there is the additional feature already noted that the soap-film surface also curves in a vertical plane.

Models can be constructed to represent other (hypothetical) planets. We can show that if there is a rapid decrease of velocity with depth, then there is bending of the waves outwards from the centre and with a low-velocity core it can prove impossible to get the soap film, and hence the waves, to penetrate.

7.6 CURVATURE OF SPACE—THE SCHWARZSCHILD GEOMETRY

Suppose we are small creatures existing within one of the two-dimensional soap films. Then we will be confined to travelling inside a two-dimensional surface, although this surface can be curved within overall space. This surface is subject to the condition that the mean curvature is zero, that is the two principal radii of curvature are everywhere equal and opposite. For the special case when the film is flat, this corresponds to zero curvature in all directions.

According to the Schwarzschild geometry of the universe (named after Karl Schwarzschild, a versatile German physicist and mathematician who lived 1873–1916), space exhibits a curvature. This is analogous to what we see in soap films, where there is curvature in two principal directions within the two-dimensional surface. At any point within the soap-film

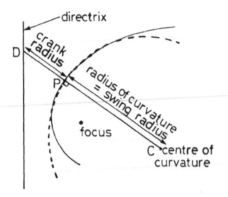

Figure 7.12. Geometry of a parabola.

surface, it is as if there are two one-dimensional lines which exhibit balanced curvature. In space in the universe, there are three dimensions and we have the balanced curvature of six two-dimensional surfaces. It is not very easy to comprehend this non-mathematically, but lines in the soap film have become surfaces, and a surface has become three-dimensional space. Thus, whereas originally we had a two-dimensional soap film which we could consider as embedded in three-dimensional space, we now have three-dimensional space which we might imagine embedded in a Euclidean four-dimensional space.

Another difference between Schwarzschild geometry and the geometry of the soap film, even if we break down our analysis to restricted components of space, is the relationship between the principal radii of curvature. It turns out that Schwarzschild curved space, when restricted to one dimension less than full space, corresponds to the geometry of a parabola. (See Wheeler 1990 for a discussion of this and much else.)

If we consider a parabola and rotate it about its directrix, then we obtain a paraboloid. This has a surface analogous to that of the catenoid film obtained between two hoops, but with a different equation to represent the different form of curve that has been rotated. The curvature at any point on the parabola has an interesting and relevant property (Wheeler (1990, p 131)). Draw the perpendicular to any point P on the parabola and obtain the centre of curvature C on this curve. This can be done by trial and error with an actual parabola and a pair of compasses. The radius of curvature PC (see figure 7.12) equals twice the distance PD as measured along the perpendicular in the opposite direction to the directrix. We can call PC the swing radius and PD the crank radius. The surface curvature will be

$$-\frac{1}{\text{crank radius} \times \text{swing radius}}.$$

(Compare this expression with that for curvature on page 10.)

The curvature in the Schwarzschild space geometry is caused by the presence of mass. The larger the mass near to the point under consideration in space, the greater the curvature. So all three principal radii of curvature in Schwarzschild space depend on the magnitude of the mass m at the centre of attraction and on the distance R between the mass and the point of observation. R is called the Schwarzschild radial coordinate. It turns out that:

$$\text{swing radius} = \sqrt{\frac{2 \times R^2}{m}} \qquad \text{crank radius} = \sqrt{\frac{R^2}{2 \times m}}.$$

Notice that if curvatures κ_1 and κ_2 correspond to the swing and crank radii respectively (where curvature is the reciprocal of radius), the sum $\kappa = \kappa_1 + \kappa_2$ is not equal to zero, so the mean curvature is not zero. Also,

$$\text{surface curvature} = -\frac{m}{R^2}.$$

One principal curvature is twice the other and this is a very important difference from the situation for soap films.

If we exist in this Schwarzschild space, we shall lie on the surface of the rotated parabola and will, of course, never encounter the directrix. We can tell that our space is curved if we and a friend travel side by side along geodesic lines (lines which are parallel at the point under consideration). We shall either gradually converge or diverge with time depending on the curvature of that part of space where we walk. This is illustrated if we travel North side by side on the earth's surface along lines of longitude starting from the equator. Whereas on starting our journey we walk precisely parallel, nevertheless we end up by meeting at the North Pole because of the curvature of the surface of the earth.

However, the sum of all three curvatures at any point in Schwarzschild space is zero. This requires further argument, which will not be considered here.

7.7 BENDING OF LIGHT NEAR A BLACK HOLE

A consequence of the curvature of space is that light travelling through curved space will follow a curved path. The *closer* the light passes to a large mass, the greater is the curvature of its path. Also, the *larger* the mass the greater the curvature. Such a curvature of the path of light will be particularly noticeable should the light pass close to a black hole. This could be the basis of identifying the presence of a black hole.

It is perhaps not surprising that the angle of deviation θ experienced by the light is inversely proportional to the shortest distance of approach l of the light to the centre of the mass (see, for instance, Misner *et al* (1973, p 184)). Using Schwarzschild geometry it can be shown that

$$\theta = 4M/l.$$

We can illustrate the bending of light using a cone-shaped perspex insert between parallel perspex plates. The perspex

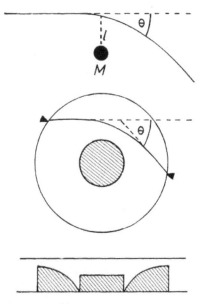

Figure 7.13. Soap-film analogue for bending of light in a gravitational field.

insert has a decreasing thickness towards the centre (figure 7.13). At the very centre we have the representation of the large mass M. The perspex insert is cut as a cone whose thickness increases in proportion to the logarithm of the radial distance. We can then see how a soap film bends as it crosses the cone-shaped region. We should remember that the deflection of the film has been modelled for one plane only whereas the bending of light in Schwarzschild space occurs in three planes.

8

Bubbles

Bubbles in fixed and variable frameworks, coalescing
bubbles and bubbles in a froth are described. Analogy
is made between patterns within a bubble raft and crystal
structures.

8.1 SIMPLE BUBBLES

We have introduced the topic of bubbles previously without
discussing their formation in detail. Soap films minimize their
film area if atmospheric pressure exists on either side. A soap
bubble established between parallel plates wants to minimize
the area it encloses. A soap bubble in three dimensions
tries to minimize the enclosed volume. Thus a single soap
bubble constrained between parallel plates will be very precisely
cylindrical (figure 8.1). The excess pressure in the bubble will
maintain the curvature in the plane of the plates and prevent it
totally collapsing. There will be no curvature perpendicular to
the plates. The film meets the plates at right angles and uniform
surface tension is maintained between the plates. As we blow
extra air into the bubble via a wet drinking straw, the diameter
of the bubble will increase. As we draw air out, the diameter
of the bubble will decrease. The *total* energy arising from the
surface energy of the film and the excess energy of the trapped
air is minimized.

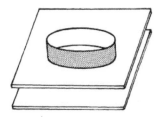

Figure 8.1. Cylindrical soap
bubble between parallel plates.

We can trap the bubble by pins between the plates. Suppose
such a bubble is trapped by three pins set at the corners of
an equilateral triangle. There will be a particular value of the
excess pressure within the bubble that will maintain the film as
a cylinder intersecting the three pins. The radius of the circular
cross-section will be the circumradius of the triangle. Increase
the pressure within the bubble and the film will bulge outwards
with a reduced radius of curvature (figure 8.2). Decrease the
pressure within the bubble and the film pulls back with an
increased radius of curvature until pairs of components meet

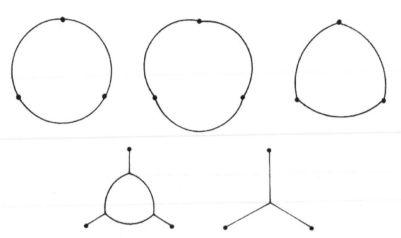

Figure 8.2   Effect of trapping a bubble with three pins and the subsequent change of pattern arising from changing the internal pressure within the bubble.

each other at 120° angles at the pins. Further decreasing the pressure produces a significant change of pattern. The size of the bubble decreases and is joined to the pins by three arms. The change to this configuration is analogous to a second-order phase change. Decreasing the pressure even further means that ultimately the bubble disappears so that the film consists only of the three linear components meeting at 120°. This change is analogous to a first-order change as there is a very slight energy discontinuity at the change. Notice that throughout the pattern changes, the film components always meet at 120° even when these components are part of the bubble surface.

Note that the film pattern shown with a small bubble of air trapped is different to the film pattern which can arise from drainage at an intersection. If a quantity of solution is retained at a junction, the so-called Plateau border is produced whereby the film now curves inwards at the junction. Such a situation is likely to be observed on a small scale at any three-component intersection (figure 8.3). Under these circumstances, only one film surface is involved around the small region of intersection. The pressure in the solution must be lower than atmospheric pressure in order to establish the opposite curvature to that observed for a trapped bubble.

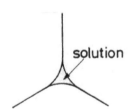

Figure 8.3.   Plateau border at a soap-film intersection.

8.2   BUBBLES WITHIN FIXED FRAMEWORKS

Often bubbles are accidentally trapped within three-dimensional frameworks when attempting to obtain film patterns. But it can be instructive to insert an air bubble intentionally. One way to achieve this is to dip the frame into the solution once in order to establish the basic film. The frame is immediately reinserted and withdrawn a second time. This traps a quantity of air. How much air is trapped will depend on the depth of submersion of the frame when dipped this second time, and on the orientation with which the frame is inserted. A little trial-and-error soon establishes the optimum angle and depth of dipping. Trapping the air produces the bubble. An alternative way to insert a bubble is to blow through a wetted straw. Again trial-and-error will show the optimum positioning of the straw before blowing and will also show the amount of air to blow into the bubble. The end of the straw should be in proximity with an intersection point of components of the film surface. Sucking out air or blowing extra air into the trapped bubbles can produce considerable variation. Surfaces always meet at 120° and lines of intersection between surfaces always meet up at 109°28′.

Usually a bubble establishes itself at the centre of the frame, and is distorted such that it shows similar symmetry to the external frame (figure 8.4). It is attached to the edges of the frame by surfaces which link with the bubble. These give it the distorted shape. Thus a bubble inserted within a tetrahedral frame (figure 8.4(a)) takes on a *tetrahedral shape* but with highly curved surfaces. These curved surfaces are maintained by the presence of an excess pressure within the bubble. As in the two-dimensional cases already discussed, the precise size of the bubble will be determined by the amount of trapped air and the need to have just the correct excess pressure to establish the necessary curvature. Even with the presence of a bubble, the component surfaces of the film always meet at 120° and the intersecting edges always meet as groups of four at angles of 109°28′. Therefore, the surfaces must curve so as to establish these conditions, but the surfaces will not be spherical. The energy minimization that the film undertakes is more complicated now that there is a bubble present. First there is the free energy which is associated with the film. This is proportional to the total area, including the area of the film that constitutes the bubble and the area of the film that makes up the side components. Secondly, there is an energy contribution from the air that is trapped within the bubble at a pressure in excess of atmospheric pressure. It is the sum of these two energy components that must be minimized in order to establish the overall shape of the film.

The bubble set up within a cubic framework is ‘cubic’ in shape (figure 8.4(c)), as one would expect by now. The sides

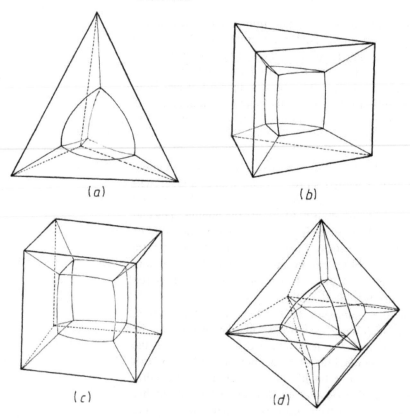

Figure 8.4 Bubbles established within regular frameworks: (a) tetrahedron; (b) triangular prism; (c) cube; and (d) octahedron.

are curved and again are joined to the edges of the frame by film surfaces very similar to those set up within the framework when no bubble is enclosed. It is possible to establish a similar 'cubic' bubble within a cluster of six bubbles. This is a technique used by a television entertainer from the USA, Tom Noddy. The six outer bubbles containing the 'cubic' bubble are equivalent to the six sides of the framework. Noddy further enhances the effect by blowing smoke into the central 'cubic' bubble to give it a whitish coloration.

It is interesting to consider the effect of sucking out the air from inside the bubble. At the point at which the bubble becomes vanishingly small there would be twelve planes meeting at a point. As we already know, this does not correspond to stable equilibrium; the film will move into the configuration with a central 'square' component in one of three possible orthogonal planes. It is a puzzling situation. Where there are curved surfaces, film patterns usually become unstable as the vertices get

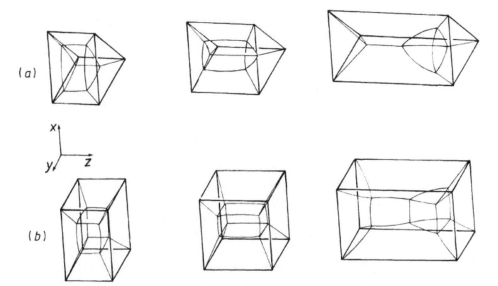

Figure 8.5 (a) Bubble within a variable triangular prismatic framework. (b) Bubble within a variable cuboidal (tetragonal) framework.

closer together, but before they meet. Here we have trapped air preventing this happening. Movement and change will depend critically on the actual air extraction, the surface tension of the film, and any movement that might cause bursting of the bubble.

8.3 BUBBLES WITHIN VARIABLE FRAMEWORKS

Bubbles set up within variable frameworks demonstrate interesting features which relate to a number of topics in this book. It is instructive to establish a bubble within say a triangular prism when its length is small and then to lengthen the framework. The bubble itself elongates and its appearance can change considerably. Its faces exhibit considerable variation of surface curvature. Eventually a framework length is reached where the bubble is no longer stable as a triangular prism with curved faces. It pulls back to form a tetrahedral bubble which is off-centre (figure 8.5(a)).

Once more, we see the analogy with a phase transition. The film patterns either side of the change have different symmetries. Assuming that the variable length of the prism is taken as the z direction, the film pattern starts off with a mirror plane in the xy plane at mid-z position, but loses this mirror plane when the framework is opened out and the tetrahedral bubble is formed.

Closing up the framework can return the bubble to its (triangular) prismatic form and the film pattern to its former symmetry. The transformation exhibits hysteresis; the changes occur at different settings of the framework as the framework is increased and reduced in length. On closing up the framework, the conversion to the prismatic form occurs when the apex of the tetrahedral bubble lying on the central axis of the film meets up with the film junction further along this central axis. On the other hand, on opening up the framework, the disappearance of the prismatic shape depends on the resulting curvature of the surfaces, and the fact that angular conditions between faces and intersecting lines must be maintained.

If the framework is closed up such that the end faces become much closer than when the bubble was first inserted, a further transition is likely to occur. Usually the bubble is pushed off-centre to one of the corners, but precisely what happens depends on the quantity of trapped air.

Next consider the cuboidal (or tetragonal) framework. When the cuboid is extended sufficiently, such that the cuboidal bubble is no longer stable, the film 'breaks' to form the usual rectangular central component of the film together with a reformed bubble which usually takes the shape of a triangular prism. One of the parallel edges of the prism also forms one of the shorter edges of the rectangle (figure 8.5(b)). The film pattern retains one mirror plane. However, if the amount of trapped air is small, as measured in terms of the overall size of the framework, a tetrahedral bubble may form, away from the central axis and at one of the corners of the inner rectangle. In this case, the overall mirror plane symmetry of the soap film is not retained.

Thirdly, we consider the pentagonal prism. When it is lengthened, the pentagonal central bubble breaks into two rectangular components but with a bubble at one end. The bubble can be a triangular prism having one edge common with an edge of one of the two rectangles (analogous to the cuboidal case). Or it can be a small tetrahedral bubble at one of the corners of the rectangles. The first case occurs with a large initiating pentagonal bubble and the second case occurs when the initiating bubble is small. In either case, no mirror plane remains after the change of pattern, except for the special case when the tetrahedral bubble happens to form at one of the two corners common to both rectangular components. Although there are six corners where the bubble might settle, the bubble will tend to go under gravity to the lowest corner, as determined by how the framework is held.

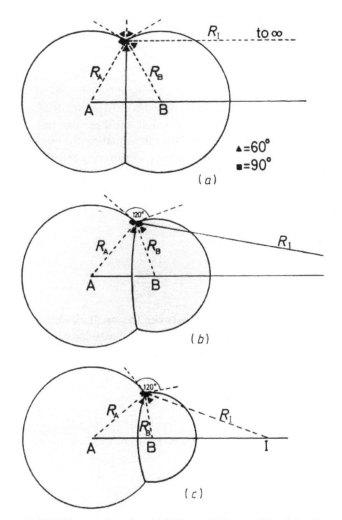

Figure 8.6 Two coalescing bubbles with soap-film interface: (a) bubbles of equal radii; (b) and (c) bubbles of unequal radii.

More complicated bubble patterns can be blown into the frameworks. Considerable variation of size and shape of bubble is possible within the variable frameworks, which provide an interesting method of looking at the divisioning and coalescing of bubbles.

8.4 COALESCING BUBBLES

The excess pressure in a spherical bubble, radius R, with two film surfaces and surface tension γ for each surface, is given by

$$P = \frac{4\gamma}{R}.$$

Suppose we have two coalescing bubbles of radii R_A and R_B with a soap-film boundary between. Because there exists an intermediate boundary, it is possible for these two bubbles to be in equilibrium. Figure 8.6 shows the situation for two bubbles of equal radius, and also for bubbles of differing radii. The excess pressure in bubble A relative to atmospheric pressure must be

$$P_A = \frac{4\gamma}{R_A}$$

and in bubble B will be

$$P_B = \frac{4\gamma}{R_B}.$$

The difference in pressure between the smaller bubble B and the larger bubble A will be given by the excess pressure P_I arising from the radius R_I of the surface of intersection

$$P_I = \frac{4\gamma}{R_I}.$$

Hence,

$$P_B = P_A + P_I$$

so that

$$\frac{1}{R_B} = \frac{1}{R_A} + \frac{1}{R_I}.$$

As commonsense would suggest, the intermediate (or partition-ing) film bends into the region of lower pressure. This is a region within the larger bubble; the larger radius of curvature must correspond to a smaller pressure. The larger the difference in size of the bubbles, the greater the curvature of the partitioning film. If the bubbles are of equal radius, there is no difference in pressure and the intermediate film surface is flat.

The film components, that is the outer surfaces of the two bubbles and the intermediate partitioning film, must all meet at 120°. This means that the tangents to the curved (spherical) surfaces at the line of intersection meet at this angle. In addition, the tangents to the spherical surfaces are at right angles to the radii of these spherical surfaces at the intersection. As a consequence, the angle between the radii of the larger and

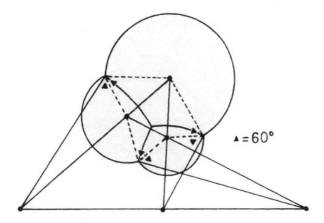

Figure 8.7 Three coalescing bubbles.

smaller bubbles, and also the angle between the radii of the smaller bubble and the intermediate surface both equal 60° at the intersections. These 60° angles are shown by triangular symbols in figure 8.6.

We can go on to consider three coalescing bubbles. There is interesting geometry associated with their arrangement, including the fact that the centres of curvature of the three intermediate surfaces all lie on a straight line. (For a discussion of this geometry see, for instance, Isenberg (1978, p 74).) A number of 60° angles are again established, and these are marked in figure 8.7. The geometry of three bubbles coalescing is thus totally prescribed. The geometry of such a grouping can be extended to larger clusters.

We saw that when two bubbles of equal diameter coalesce, the intermediate surface is a plane with infinite radius of curvature. It is very easy to see what happens for larger clusters of bubbles by blowing bubbles onto a flat perspex surface (even though the angle of contact will not be 90°). The geometry is interesting for clusters of bubbles having equal radii. Examples of such clusters are shown in figure 8.8. Because radii are equal, the pressure within each separate bubble will be identical, and hence each partitioning film component is planar. These planes establish similar patterns to those obtained for films joining pins between parallel plates. We should note that in the case of three bubbles, there are three equi-lengthed planar surfaces intersecting at 120° to each other. This is exactly analogous to the film formed between three pins arranged at the corners of an equilateral triangle. For the case of four bubbles, the planar surfaces do not extend to the corners of a square, rather they

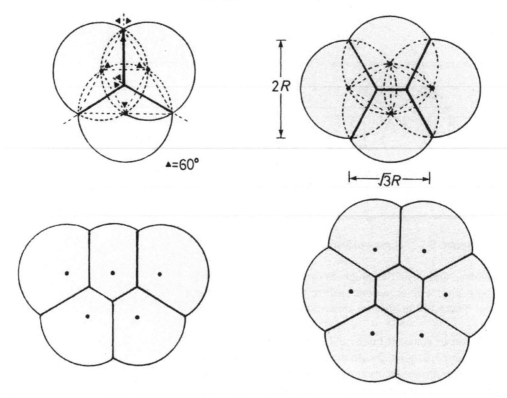

$\blacktriangle = 60°$

Figure 8.8 Multiple coalescing bubbles of equal radii.

extend to the corners of a rectangle whose sides have lengths in the ratio $2:\sqrt{3}$.

Five bubbles give a similar series of film components to those of the five-pin film pattern, but again the lengths of the components are different. A seven-bubble pattern can be produced in which the central bubble has six sides forming a hexagonal pattern. All these bubbles including the central bubble will have the same curvature. One might think that the central hexagon can be made any size without changing its internal pressure. Adding or subtracting air merely changes the size of the dome of the hexagon without changing its curvature. Nevertheless this will alter the shape of the configuration slightly as this dome must meet the surfaces of the surrounding bubbles at precisely 120°. Such variations are difficult to detect by eye.

The coalescence of bubbles of unequal radii can be used for molecular modelling. Rämme (1992) has suggested the modelling of molecules such as H_2, H_2O, HCl and F_2 by producing bubbles with diameters of size in proportion to the single-bond covalent radii of the individual atoms.

The patterning changes somewhat if bubbles are arranged in three-dimensional space. Although it is possible to establish experimentally three-dimensional bubble patterns, for instance on the end of a wetted drinking straw, the study of these is more difficult and such bubble patterns will not be considered here.

Obviously if the bubbles are just allowed to touch without coalescing, the spherical shape of the bubbles is retained. A honeycomb lattice of bubbles can be produced and the bubble raft which is made up of such an arrangement of bubbles will be considered later in the chapter. In the case of coalescing bubbles, a hexagonal pattern can arise from zero pressure difference between the bubble cells and the effect of surface tension within the film. In the case of bubbles on a bubble raft, the hexagonal patterning arises from excess pressure within the bubbles and the fact that the bubbles close-pack together.

8.5 MORE ABOUT BUBBLE PATTERNS

We can set up soap cells between parallel perspex plates to produce extended patterns. Narrower spacing between the parallel plates than the diameters of the bubbles is needed for this; i.e. we need a method of producing bubbles via an orifice of a suitable size such that we see not complete spherical bubbles but squashed polygonal cells.

The bubbles can grow or shrink over an extended period of time by the diffusion of gas from nearest neighbours, such that a dynamic system evolves. As a result, the rate of change of volume of a bubble cell (equivalent to the rate of change of area of a cell as seen from above because of the constant height of the cells) depends only on the number of vertical faces. This is referred to as Von Neumann's law of bubble growth (Von Neumann (1952, p 108)). It can be argued that a strictly regular array of hexagonal bubbles should be indefinitely stable. These arguments lead to an equation for the rate of change of area of the form

$$\frac{dA}{dt} = K(n - 6)$$

where A is the projected area of the bubble, t is time, n is the number of sides and K is a constant related to the rate of diffusion of the gas. Strictly in obtaining such an expression, polygons with $120°$ internal angles are assumed. This is not completely true in practice. With all angles $120°$, there will be finite curvature of the sides of the polygons other than six-sided

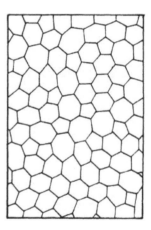

Figure 8.9. A typical two-dimensional cellular film pattern.

Box 8.1   Mathematics of two-dimensional cellular structures

In two dimensions, the cellular structures (figure 8.9) obey the conservation law given by Euler's equation (see Weaire and Rivier 1984)

$$F - E + V = \chi$$

where F is number of faces, E is the number of edges, and V the number of vertices. χ is 1 for a plane (2 for a sphere). The average number of edges to a cell is $\langle n \rangle = 6$. Three edges meet at a vertex and an edge links two vertices. As a consequence we have the valence relations

$$3V = 2E \qquad \text{and} \qquad \sum_n n F_n = 2E$$

where F_n is the number of n-sided cells. So Euler's relation becomes

$$6 - \frac{1}{F} \sum_n n F_n \approx 1$$

and for a large number of cells

$$6F \approx \sum_n n F_n.$$

In three dimensions, the equivalent Euler equation is

$$-C + F - E + V = \xi$$

where C is the number of cells. Vertices have a coordination of 4 instead of 3 and the average number of faces $\langle f \rangle$ is given by

$$\langle f \rangle = 12/(6 - \langle n \rangle).$$

Most random structures such as foams have an average number of faces slightly less than 14. However, this is not a limiting value for $\langle f \rangle$. The average number of vertices is approximately 23 and the average number of sides to each face is fractionally over 5.

ones. This curvature is usually not shown in diagrams of such bubble patterns.

As patterns alter with time, there is a rearrangement of the cells. A process in which the vertices rearrange (illustrative of a first-order phase change) is called a T1 process. Alternatively, faces may disappear and as a consequence two-dimensional cells may disappear. Such a process is a T2 process and is analogous (at least approximately) to a second-order phase change. The T1 and T2 processes for a three-sided cell are shown in figure 8.10. Cells with more than three sides can disappear also, but these disappearances can be reduced to a combination of T1 processes plus a T2 process. Division of cells can be explained similarly by a combination of T1 and T2 processes, but in reverse. However, divisioning tends to be sudden rather than a succession of processes, and is illustrated in biological systems, where sudden divisioning is called *mitosis*.

The evolution of soap froths as contained between parallel perspex plates and the distribution of the number of sides of bubbles has been investigated by Stavens and Glazier (1989). The number of bubbles was found to decrease with time starting with approximately 10,000 bubbles and ending with approximately 100. The patterning was initially built up largely of hexagonal bubbles exhibiting a uniformity analogous to a crystalline grain, but with some five- and seven-sided bubble cells present producing boundaries to these grains. (See also the discussion in the next section concerning the analogy between bubble patterns in a bubble raft and crystalline structures.) The distribution of numbers of sides was obtained for the limit of long time of evolution. Five-sided bubbles were most common; but six-sided were nearly as common. The distribution fell away either side of these values. There were considerably less bubbles having four sides and a negligible number having three. A negligible number of bubbles possessed nine sides and above (figure 8.11). This limiting distribution appeared to be independent of the starting size of the bubbles. Deviations of angles from the precise angle of 120° were identified, and this implies some deviation from Von Neumann's law.

Cellular patterns similar to soap-film patterns are observed in uniaxial magnetic garnet films. The ordered hexagonal domains can show fronts advancing in a similar manner to the melting of crystals (Babcock and Westervelt 1989). Hence ordered and disordered phases can be shown existing alongside each other. The control parameter is the magnetic field and a change of biassing field can produce phase transitions and changes in cell

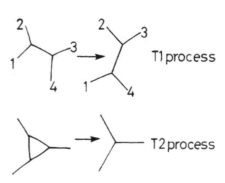

Figure 8.10. T1 and T2 processes.

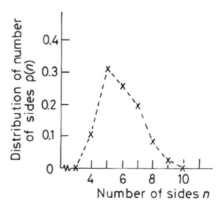

Figure 8.11. Distribution of the number of sides $\rho(n)$ of bubbles in a froth (from Stavans and Glazier 1989).

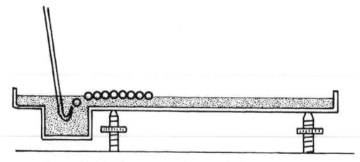

Figure 8.12   Apparatus for creating a bubble raft.

density. A plot of cell density versus magnetic field can produce
a form of phase-transition diagram in which there is a jump in the
cell density. There is a higher cell density in the regular (lattice)
phase and a lower cell density in the phase which appears rather
like a froth.

8.6   THE BUBBLE RAFT AND CRYSTAL STRUCTURE

The development of bubbles in a bubble raft was studied
extensively by Bragg and Nye (1947) in order to create a
dynamical model of a crystal structure, notably the crystalline
structure of a metal. As we shall see, the arrangement of the
atoms within a metal can often be considered as the packing of
a large number of spheres, and this is analogous to the packing
of a large number of bubbles. The raft model produces packing
within two dimensions.

Apparatus suitable for creating a raft of bubbles similar to
that used by Bragg and Nye is shown in figure 8.12. It is
advantageous to blacken the bottom of the apparatus tray so that
the bubbles and their arrangement show up clearly. The bubbles
are blown from a fine orifice beneath the surface of the soap
solution so that the bubbles can rise and float on the surface. With
the orifice of the jet about 5 mm below the liquid surface and
with a small constant air pressure, bubbles of remarkably uniform
diameter are produced. If the formation of bubbles becomes
irregular, unwanted bubbles can be destroyed by playing a small
flame over the region of irregularity on the surface. Bubbles of
diameter 2.0 mm and less can be produced, although very small
bubbles must be produced by an alternative technique involving a
rapidly rotating dish. The bubble diameter varies with orifice size
but not significantly with pressure or depth of the orifice below
the surface of the liquid. The rate of production of bubbles is,

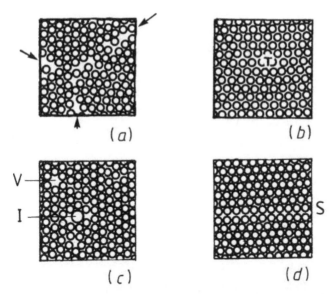

Figure 8.13 Diagrams of bubble arrangements illustrating crystal defects: (a) grain boundaries; (b) dislocation; (c) impurity atom and a vacancy; and (d) fault line.

however, very dependent on the excess pressure applied at the orifice. Bragg and Nye found that a thin-walled jet of 27 μm diameter produced bubbles of approximately 0.6 mm.

Bragg and Nye used assemblies of bubbles numbering up to the order of one hundred thousand. The bubbles are attracted together by a capillary type interaction, analogous to the metallic bonding arising from the free electrons in a metal. Rafts of bubbles of great regularity can be produced, but interest centres particularly on the observed irregularities, which are analogous to crystal imperfections. Various arrangements of bubbles that arise are illustrated in figure 8.13.

For instance, it is possible to observe separate regions of regular packing which are oriented differently, with the regions between consisting of bubbles that are slightly disturbed with respect to those nearby. However, the bubbles tend to adhere to one region of regularity or the other. The regions of regularity are equivalent to grains and the intermediate regions to grain boundaries (marked with arrows in figure 8.13(a)). Bubbles which are slightly larger or slightly smaller than the norm tend to aggregate at the grain boundaries. However, the bubbles are not able to diffuse through the raft. This is different to the situation in a crystal where atoms are able to diffuse past each other. So

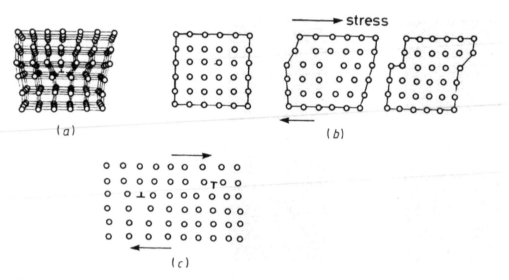

Figure 8.14   (a) Edge dislocation in a crystal. (b) Movement of an edge dislocation under the application of a force. (c) Annihilation of positive and negative dislocations on coming together.

the fact that these 'impurity' bubbles tend to form at the grain boundaries arises because of the way the grains develop at the expense of neighbouring ones, and changes of size tend to stop when the boundary reaches an 'impurity' bubble.

The bubble raft is especially useful for demonstrating dislocations (position marked T in figure 8.13(b)), both their presence and possible movement. If the raft is compressed or otherwise deformed, what happens can be similar to the deformation of a metal. There is a small amount of movement as the model contracts within its elastic range. But beyond a certain point of compression, slippage of bubbles occurs, one region slipping relative to another. This slip will be along one of the three equally inclined directions of the close-packed rows. Slippage of one row over another will occur in stages, each by the distance separating nearest neighbours.

The dislocation itself is the presence of an extra bubble in the row on one side of the slip line in the raft compared to within the row which exists on the other side. This is analogous to an extra plane of atoms in a crystal lattice (figure 8.14(a)). The dislocation, that is the presence of the extra bubble, can move along the slip line when induced to do so by a small force exerted on part of the raft. It is the net result of this that produces the slip of one portion of the bubble raft by one atomic or one bubble spacing. The mechanism is used to explain plastic deformation of crystals by quite small forces; this is because dislocations in crystals can move in the same way, one crystal spacing at a time, under the influence of a force (figure 8.14(b)). The dislocations can be seen to move very slowly or quite quickly across the raft. The length of dislocation depends on the nature of the strain. On the bubble raft, the greater the rigidity of the bubbles, the longer will be the dislocations (as measured in terms of the number of bubbles). More rigidity is obtained when the bubbles are of smaller size.

As in real crystals, we can have bubble dislocations which are positive or negative according to whether the extra bubble (or bubbles) is in the row (or rows) above or below the dislocation. When a positive and a negative dislocation come together, they can be seen to annihilate each other. The same effect occurs in real crystals (figure 8.14(c)).

If the bubble raft is gently agitated it is possible for a hole (i.e. vacancy) to appear. A number of vacancies always occur in a crystal structure; it is not thermodynamically possible to have a perfect crystal with no vacancies. The appearance and also disappearance of bubble vacancies on the raft, whilst it is gently agitated, is analogous to the appearance and disappearance of vacancies in a crystal as it is 'cold worked'.

The effect of an impurity atom can be seen by intentionally inserting a bubble of differing size from the others. This will disturb the regularity of the lattice. Strain is created in the bubble raft in the same way that strain, and hence strain energy, is incorporated into a real crystal. If the 'impurity' bubble arises accidentally in the formation of the raft, then it tends to get caught up by a grain boundary. The formation of an impurity atom or bubble (I) and also the presence of a vacancy (V) are shown in figure 8.13(c).

Figure 8.13(d) shows a narrow strip (S) between bubbles or crystals that exhibit parallel orientation. It is as if this region is crossed by numerous fault lines. Recrystallization might be expected to occur here and the region of perfect crystallinity gradually to increase in size.

If a raft of bubbles is given a vigorous stirring it gets broken up into a number of separate crystalline regions or 'crystallites'. Left to readjust, the raft tends to recrystallize into larger grains. This is similar to annealing of crystal material when held at an appropriate temperature, usually above ambient temperature.

An alternative model for crystalline materials, that operates in a similar manner, consists of small ball bearings held between glass plates. (It is as if we were to incorporate ball bearings within a projector slide.) When the plates are tapped, the ball bearings move into regions of regularity. These correspond to crystalline regions, each separate region constituting an individual grain separated from the other regions by grain boundaries. Once more it is possible to see dislocations and vacancies, and to see their rearrangement as the model is tapped. Suitable mechanical shaking of the slide can produce agitation corresponding to thermal agitation. The interface between filled and unfilled regions of the slide becomes less well-defined. The ball bearings at the interface move around, and the crystal (the array of ball bearings) starts to 'melt' or even 'vaporize'.

9

Analogies within the scientific world

A varied selection of analogies from areas of biology, chemistry, physics and technology are chosen for discussion. Topics include radiolarians, fullerenes, the honeycomb, crystal habit, Langmuir–Blodgett films, communication networks and architectural structures.

9.1 INTRODUCTION

It has been shown in the earlier chapters of this book that soap films attempt to minimize their energy and hence their area. Also, because minimization principles underpin so much of the scientific world, both physical and biological, the formations of soap films can be used to demonstrate and explain very disparate phenomena. Many examples have been discussed already. Collected together in this last chapter are a variety of further scientific examples in which useful parallels can be made with our soap-film studies. All the examples involve patterns, shapes, or structural form. Underlying forces or underlying energy considerations have given rise to the actual structural forms.

9.2 BIOLOGICAL UNICELLULAR FORMS AND RADIOLARIANS

We have seen the analogy between the existence of simple cellular structures in nature and bubble formations. Unicellular biological beings are essentially drops of protoplasm. If formed by suspension within water, they might be expected to be spherical in shape. Many are, but others are not. The spherical ones usually show very little motion. Non-spherical ones tend to have protruding cilia or flagelli (long cilia), which are whip-shaped protrusions that can wave around and produce mobility.

D'Arcy Wentworth Thompson (1860–1948), in his classic book *On Growth and Form* (1942, first edition 1917), notes unicellular forms which exhibit all types of H-surface (see page 81). The cilia or flagelli are important in the stability of these non-spherical organisms. Although surface tension clearly plays a part, there is a membrane tension which contributes. Both tensions act very similarly, although membrane contraction does not occur beyond a certain limit and so the establishment of equilibrium can differ somewhat.

Radiolarians show remarkable parallels with the film and bubble patterns produced in frameworks as referred to in Chapters 6 and 8. Many radiolarians have structures, usually made of silica, that are similar to cubic, triangular-prismatic, tetrahedral and other frameworks.

Ernst Haeckel (1834–1919) went on the Challenger expedition of 1873–6 and produced some 4700 drawings of radiolarians that he had studied under the microscope. These microscopic organisms were some of the earliest living things on earth, and as such can be expected to reflect rather simple methods of production. Thompson drew the analogy between their forms and the shapes of soap bubbles. He discussed the surface energy conditions which may lead to their production. However, the more recent use of electron microscopes has tracked down a high level of complexity on a very fine scale, including the presence of large numbers of elastic, but totally straight, threads called axopodes that are fundamental to the metabolism of the radiolarians. There are also very fine tubes called microtubuli arranged in highly regular arrays, for instance hexagonal arrays, and these microtubuli are connected such that matter can be transported.

The radiolarians tend to consist of protoplasm, sometimes surrounded by a froth of cells, but usually there is a distinct separation between protoplasm and water within the defined volume of the radiolarian. Thompson considered the possibilities when two fluids are in contact, in this case protoplasm and water. A body immersed in the fluids will be either totally immersed in the protoplasm, or totally immersed in the water, or alternatively sitting at the boundary between the two. There are three interfacial contacts possible, each with their own surface energy per unit area. We denote these as γ_{wb}, γ_{pb} and γ_{wp}, where subscripts w, p and b stand for water, protoplasm and body. The surface energy when the body is totally immersed half in water and half in protoplasm is

$$(S/2)(\gamma_{wb} + \gamma_{pb}) - A\gamma_{wp}$$

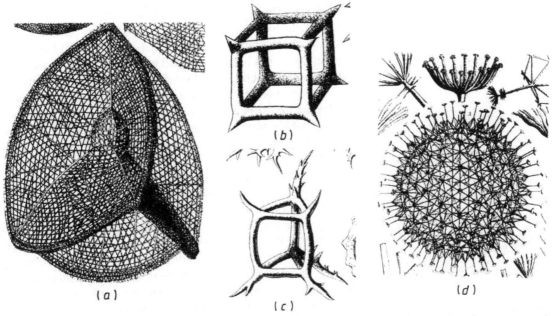

Figure 9.1 Examples of radiolarians: (a) *Callimitra agnesae*; (b) *Lithocubus geometricus*; (c) *Prismatium tripodium*; and (d) *Auloscena mirabilis*.

where S is the surface area of the body and A the cross-sectional area. The last term (a negative correction term) arises from a consequent release of energy. Totally immersed in water or protoplasm, the energy terms will be $S\gamma_{wb}$ and $S\gamma_{pb}$ respectively. Minimal potential energy will arise when the body is approximately half in and half out of each fluid. So frameworks and specules (radiolarians which each look rather like T-shaped spikes) tend to exist in a position penetrating both protoplasm and water.

Let us consider some specific shapes of radiolarians that are similar to shapes we have already seen. One particular group of radiolarians are called *Nasellaria*, of which *Callimitra agnesae* has a skeleton of the form shown in figure 9.1(a). There is a good analogy with the soap-bubble pattern established in the tetrahedral framework (figure 8.4(a), page 134). The skeleton's diameter is approximately 0.15 mm. It should be noted that the outer framework, as well as the inner framework, has *curved* surfaces. Such a situation would arise if the outer framework, which is a spherical tetrahedron, was itself originally supported by a tetrahedral cluster of bubbles.

Another radiolarian of interest as shown in figure 9.1(b) is *Lithocubus geometricus*. (Note the name attributed to it by

Haeckel.) This radiolarian is similar to the cubic framework used for demonstrations in Chapter 6, but it has probably arisen from a bubble arrangement in which a central 'cubic' bubble (see figure 8.4(c)) has been established by a surrounding arrangement of six other bubbles. The skeleton may well establish itself at the surface boundaries of the bubbles. A further framework is *Prismatium tripodium* (figure 9.1(c)) analogous to the shape of the bubble formed within the triangular prism. Clearly we can expect (and do find) additional frameworks corresponding to other bubble arrangements. It is easy to conceive that many complicated radiolarians can arise by more and more components solidifying out either simultaneously or consecutively.

It is possible to find radiolarian frameworks consisting of a large number of polygons built up over a spherical surface (figure 9.1(d)). The polygons consist mainly of hexagons (we have noted previously the very high stability of the hexagonal shape in soap-film arrangements), but with pentagons also present. It is easy to see the need for pentagons as well as hexagons on a curved surface by looking at the stitched or pressed-out panels on many footballs (figure 9.2).

Figure 9.2. Pattern of panels on a football.

The same arrangement of hexagons, with pentagons mixed in, is observed over the surface of the eye of the fruit fly. Such an observation can be made by viewing a fruit fly in the beam of a scanning electron microscope. However, it is necessary to first coat the fly with a thin film of metal, such as gold, so that its surface is conducting; otherwise the fly rapidly builds up charge within the electron beam. In addition to the hexagonal pattern of eye facets, there is a considerable amount of other detail over the surface of the eye, including numerous hairs (see figure 9.3).

9.3 FORMS OF CARBON AND THE FULLERENES

The patterning observed with soap films has considerable analogy with molecular and crystal structures and in particular with the molecular arrangements of carbon. Thus the stable hexagonal shape as produced by soap films joining six hexagonally placed pins between parallel perspex plates, or as often produced as part of the soap-film pattern within wire frameworks, is analogous both to the hexagonal carbon arrangement in the molecular layers in graphite and to the carbon arrangement in the benzene ring (figures 9.4(a) and (b)). We have already noted that the intersection of four film edges at 109°28′ angles is analogous to the molecular arrangement of the carbon atoms in diamond

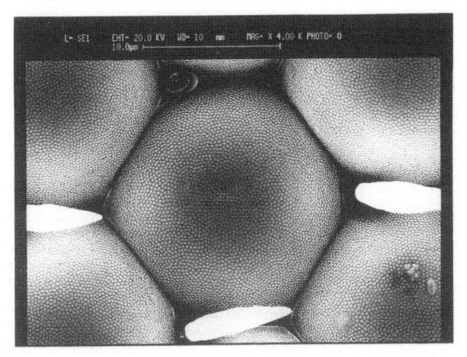

Figure 9.3 Photograph of the eye of a fruit fly obtained using a scanning electron microscope with a magnification of ×4000 (courtesy C D Meekison, Physics Department, University of Essex).

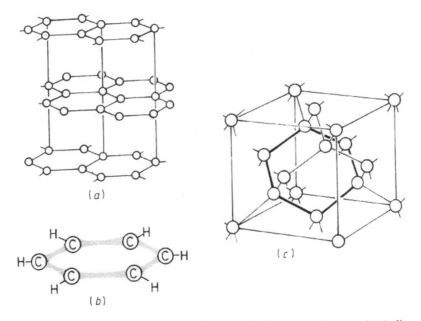

Figure 9.4 Bonding structures in: (a) graphite; (b) the benzene ring; and (c) diamond.

(figure 9.4(c)) and to the atomic configuration for zinc blende and gallium arsenide.

Although not self-supporting, the stable surfaces formed around multi-faced regular polyhedra are analogous both to the outer shapes of such radiolarians as illustrated in figure 9.1(d) and also to the fullerenes and, in particular, the highly stable C_{60} fullerene 'buckminsterfullerene'. If we count up the number of corners to the stitched panels of the football illustrated in figure 9.2, we obtain a total of 60, equal to the number of carbon atoms in buckminsterfullerene. This carbon form can be produced with considerable difficulty by techniques involving vaporization of graphite. Carbon atoms lie in the same positions in the structure as the corners of the panels of the football. Buckminsterfullerene is essentially spherical. Its name comes from the invention by the engineer R Buckminster Fuller of this geometrical arrangement as a geodesic dome in architecture. Details for reproducing graphically the gallium arsenide and buckminsterfullerine structures using *Mathematica* have been given by Mitchell and McLean (1993).

Other fullerenes can be produced with more or fewer hexagonal faces and differing proportions of pentagons (see Curl and Smalley 1991). None of these are as spherical as buckminsterfullerene. Certain ones are of pure carbon, whereas others include further elements. Buckminsterfullerene is nicknamed 'buckyball' and other arrangements are called 'buckybabies' where they have a significantly smaller number of faces, 'fuzzyballs' when the carbon atoms are hydrogenated, and so on. Coalescence of fullerenes has been reported (Yeretzian *et al* 1992) and it has been suggested by Ohmae (1993) that this coalescence, and also, the as-yet hypothetical rearrangement and splitting of fullerenes, can be modelled by coalescence, rearrangement and splitting of soap bubbles.

9.4 THE HONEYCOMB

The honeycomb as made by bees consists of cells with hexagonal cross-sections of surprising regularity. Consequently, one might deduce parallels with the 120° angles obtained between soap-film surfaces and the very stable two-dimensional hexagonal film pattern discussed in Chapter 2. Indeed we are right to make comparisons, but we should not take the parallels too far.

Bees build in a similar manner whether in the hive or in the wild. The upper part of the hive is used for honey storage and it is from this part of the hive that the familiar honeycombs are

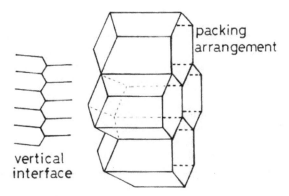

packing arrangement

vertical interface

Figure 9.5 Rhombic dodecahedra arranged as in a honeycomb.

obtained as the source of honey. In the lower part of the hive are the brood cells, which are slightly larger and have approximately the dimensions of the bodies of the female worker bees. Within these brood cells there are some even larger cells; these are for the drones, the unfertilized males bred to mate with a future queen.

The bees commence their combs from the top by attaching blobs of wax which extend down vertically. The wax is secreted from their abdomens and is transferred using the back legs of the bees from the abdomens to their mouths. Here the wax is masticated to the correct consistency for building the comb. The comb is kept at above 32 °C at which temperature the wax is a viscous fluid. Surface tension may play an important part in this process.

The cells are a series of open-ended rhombic dodecahedra (see figure 9.5) built from opposite sides of a vertical interface; the ends of the dodecahedra on opposite sides of the vertical interface are arranged alternately with each other and slightly interdigitated. In addition, they tilt very slightly upwards from the central interface. There are no gaps either between the cells on either side of the vertical interface or between cells which are adjacent. This arrangement could be due to surface tension effects, but is more likely due to the opposing pressures of adjacent worker bees as they establish the cells.

Wasps use a form of paper for their nests or combs. They chew wood or vegetation in their mouths and mould it to the shape of cells which can be as precise in cross-section as those of the bee. They do not, however, build to a vertical interface as found in the honeycomb of the bee.

The fact that both bees and wasps produce hexagonal prisms

Table 9.1 Relative efficiency of polyhedra for space-filling applications.

Polyhedron	Surface area:volume ratio
cube	6.000
hexagonal prism	5.942
elongated dodecahedron	5.470
rhombic dodecahedron	5.345
truncated octahedron (also called tetrakaidecahedron)	5.314

arises because this is a very efficient way to fill a two-dimensional plane with cells. There are only five polyhedra which can be used for space-filling. Their efficiency (or perhaps one should say inefficiency) can be judged by the ratio of surface area to fixed unit volume (Ollerenshaw 1979b) as shown in table 9.1. This table uses a convention for prisms and elongated dodecahedra in which height is taken as having the same value as the length of a regular polygonal side. The regular dodecahedron (i.e. the Platonic form having twelve equal pentagonal faces) is nearly space-filling and has a ratio of surface area to volume of 5.312. The elongated version is space-filling with a slightly increased ratio.

If one bisects a rhombic dodecahedron and a shortened truncated octahedron such that they have equal volumes and equal hexagonal openings in order that they produce honeycombs of equal width, then the latter has the smaller surface area (Fejes Tóth 1965). This will also apply for any elongation of the two polyhedra. So if the bees were to close the ends of their cells with two hexagons and two rhombi, this would be more efficient on surface area, and hence on wax, than closing the cells with three rhombi. Such cells would save 0.35% of the area of the hexagonal opening! But the cells of a comb are not regular to this degree of accuracy.

Overall, it must be significant that the dihedral angles of the rhombohedron are 120° and all edges meet at 109°28′; the achievement of area minimization occurs by one natural mechanism or another. The naming of the 109°28′ angle as the Maraldi angle arose from the measurements made on the cells in the early eighteenth century by the astronomer of that name.

9.5   CRACKS AND FISSURES

Cracking is a very complex phenomenon in nature; it is seen in mud patterns, where shrinkage of the mud has occurred during drying, it is seen in the old bark of trees, and it is seen in fissures in the earth.   Investigation of cracking indicates the frequent presence of 120° angles between cracks, but in addition 90° and 60° angles.   It has been suggested (see for instance Stevens 1976) that the particular angles between cracks depend critically on whether the cracking occurs in an elastic or an inelastic material.   For an elastic material (and rocks fit this category reasonably well) rupture occurs suddenly, and produces the 120° angle between rupture lines. For non-elastic materials it is suggested that the cracks should occur at 90° angles to each other.   In these materials, bonds between molecules do not yield as the stress builds up.   Rupture occurs at the point where the stress is initially applied.   The cracks release all the stress in a single direction parallel to themselves and perpendicular to the direction of greatest tension. Cracking occurs sequentially with succeeding cracks running and meeting at right angles.

Materials which exhibit a mix of elastic and non-elastic behaviour, possibly on two levels of size, may show both forms of cracking.   60° angles may arise as a hybrid form when one crack from an array of three has been suppressed by an irregularity in the material suppressing cracking.

9.6   EQUILIBRIUM SHAPES OF CRYSTALS

We have often made comparisons between soap-film patterns and crystal structures.   In particular, we have drawn the analogy between changes of soap-film patterns, as we alter a control parameter (usually a component of length or angle), and the change of crystal structure as a result of a variation in a thermodynamic parameter such as temperature or pressure. Crystal systems are very much more complex than the soap-film structures. The fact that they are space-filling within three dimensions, whereas soap films are two-dimensional structures even though they exist in three-dimensional space, constitutes a fundamental difference.

However, underlying both types of structure are similar principles of physics.   We can, for instance, consider the thermodynamics at the surface of a crystal or interface.   For soap films we assume the surface tension is constant over all its surface. There is no variation of surface tension with direction, and by keeping the ambient temperature uniform and pressure constant, there is no variation between different regions of the film. Consequently, when considering soap films, minimization of energy means minimization of area.

Crystal surfaces will have a surface tension associated with them, but the significant difference from soap films is that surface tension is now variable with direction. The arrangement of the atoms within planes will depend on the ordering and bonding. Both of these, and so therefore surface tension, can vary with direction. The surface tension will be of different magnitude on differently oriented faces of a crystal.

For equilibrium of a crystal surface, it is the product of surface tension and area as summed over the whole external surface which will be minimized. Or at least that will be the target when the material crystallizes, although local variation of conditions as established by the environmental circumstances may prevent this in reality. The minimization of surface energy gives rise to the distinctive facial nature of crystals.

The method of showing why the crystal faces originate as a consequence of the minimization of surface energy (for a fixed volume of material) is due to the crystallographer Georg Wulff (1863–1925). A determination of the expected pattern of faces can be obtained from a knowledge of the variation of surface tension with orientation by using the so-called Wulff construction.

A polar diagram of the variation of surface tension γ is made. In this diagram, a vector from an origin point to any point on the plot represents firstly the direction of a normal to a particular plane and secondly the magnitude of the surface tension for that plane. (This is possible since a vector has both direction and magnitude.) The diagram should be in three dimensions, but in practice we plot two-dimensional cross-sections. Estimation of the surface tension values can often be achieved by looking at the bonding pattern and the bond breakages which must have occurred at the surface. More sophisticated techniques are possible, such as computer calculations based on nearest-neighbour interactions.

Figure 9.6 shows a Wulff plot for a cubic-structured crystal, where the plot lies in a plane parallel to one of the cubic faces (see for instance Blakely (1973, p 13), based on Herring (1953)). Having drawn the plot, we draw a series of vectors P from the origin O to the plot itself. We then draw perpendiculars to each of the vectors P at the points where they intersect the γ-plot. The perpendicular planes (in three dimensions) to P are referred to as Wulff planes, and each is parallel to the crystal

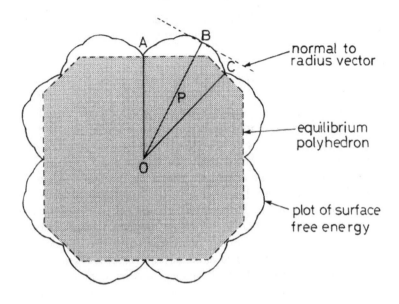

Figure 9.6 Wulff plot for cubic crystal.

surface which a particular vector P represents. We draw vectors around the entire Wulff plot together with their perpendiculars. The equilibrium shape of the crystal is determined by finding the inner envelope of all the Wulff planes, and the method is called the Wulff construction. We can see from figure 9.6 how a particular direction (A say), that has associated with it a low surface tension, can have a dominant effect in establishing a face. Another direction in the figure where the surface tension γ minimizes (C) also gives rise to a face; but at B, where there is yet a further local minimum of γ with respect to variation of angle, no face arises. Having obtained the shape of the crystal, the outline can be rescaled to produce the required volume. Any other polygonal shape will possess a greater surface energy than the one calculated, assuming we started with a correct γ-plot. Although it is possible to use a Wulff construction to turn a γ-plot into a unique plot of the equilibrium shape of the crystal, it is not possible to go from the crystal shape (called habit) to the Wulff plot.

As we might expect from the Wulff construction, crystals usually form with distinctly planar faces; that is, they are polyhedral in shape. Although surface tension plays an important role in determining the surface forms of soap films, soap bubbles *and* crystals, the external form of the crystal (crystal habit) is very different from that of the soap bubble. It is only when there are no directions in which the surface tension has a distinctly low value that we encounter spherical or ellipsoidal single crystals.

9.7 LANGMUIR–BLODGETT FILMS

So far we have been discussing free-standing films, in which soap molecules or similar surfactants form a double-film surface with a thin layer of water between. Single-sided films of one molecule thickness can be established on a water surface, the water being called the subphase. This is the basis of a scientific method of producing thin films built up from repeated layers.

Benjamin Franklin (1706–1790) carried out an experiment in 1757 on Clapham pond in which he dropped a small amount of oil onto the surface of the pond and watched it spread out. He appreciated the thin nature of the film. Over a century later, Irving Langmuir (1881–1957) spread films of monolayers onto a trough of water, and, together with his assistant Katharine Blodgett (1898–1979), investigated the transfer of monolayers from the trough to a substrate, such as a glass slide, dipped through the surface of the liquid in the trough. The classic material used for this is stearic acid ($C_{17}H_{35}COOH$), which has a hydrophilic, or water-loving, headgroup COOH that is highly soluble in water, and a long alkyl chain $C_{17}H_{35}$ which forms a hydrophobic, or water-hating, tail. When a solution of the stearic acid is made in a water-immiscible solvent such as chloroform, the solution spreads rapidly across the available area of the trough. This solvent evaporates to leave behind the stearic acid film. The similarity to the use of sodium stearate or similar fatty acid salts for soap films is obvious, and, in fact, films made up of fatty acid salts, usually ones containing cadmium, can be grown by adding a salt such as cadmium chloride to the water subphase in small quantities.

Figure 9.7(a) sketches the arrangement of a modern Langmuir–Blodgett trough. The surface area of the trough is enclosed by a continuous tape such that the confined area can be altered without altering the perimeter of the enclosed area. Figure 9.7(b) shows how the continuous tape is passed around a set of rollers, two of which move in and out to adjust the enclosed area. This variation allows the film on the surface of the water subphase to be compressed and expanded as desired without it escaping through any gap in the perimeter tape.

When the stearic acid layer is expanded over the surface, the tails of the molecules flatten out. As the end section of the barrier is brought in and the monolayer compressed, the tails

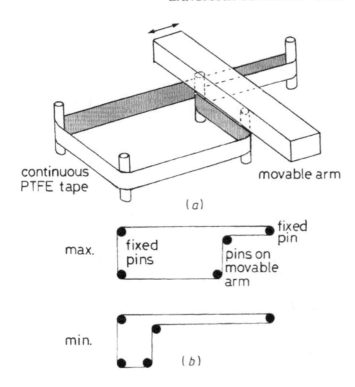

continuous
PTFE tape   movable arm

(a)

max.   fixed
            pins   pins on
                             movable
                             arm

                                          fixed
                                          pin

min.   (b)

Figure 9.7 (a) Langmuir–Blodgett trough. (b) Altering the surface area of the trough.

start to sit up, and eventually all are set vertically as shown in figure 9.8. Further compression leads to collapse of the layer. The progress of the film as represented in figure 9.8, can be demonstrated on a plot of surface pressure versus area per molecule (figure 9.9). Surface pressure is measured using a strip of filter paper inserted into the surface. Solution rises up the paper to produce a force which is proportional to the surface tension of the film. Calibration is carried out against pure water and the change of actual force measured, usually electronically, when a monolayer is added to the pure water. The value for pure water is high (73 mN m^{-1}) and is one reason why water is a useful subphase.

We see from figure 9.9 that for an expanded monolayer we start with a very low value of surface pressure and we watch this pressure increase as compression occurs. The surface pressure plot shows a number of distinct regions. Firstly, an approximately horizontal region is reached as shown in the figure inset. The hydrophobic tails are gradually lifting up

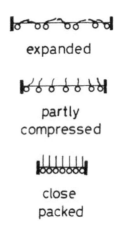

expanded

partly
compressed

close
packed

Figure 9.8. Compression of the monolayer.

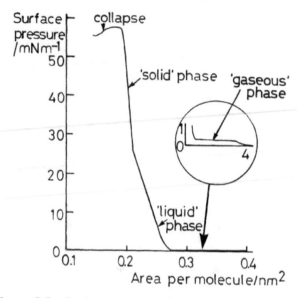

Figure 9.9   Surface pressure plotted versus area per molecule.

from the water surface. Next, there is an abrupt change to a region exhibiting considerable slope on the graph. This region represents circumstances somewhat analogous to the compression of a liquid. Suddenly, there is a further increase in gradient. Here, an ordered two-dimensional array of molecules is formed, rather like an ordered solid. Further closure of the barrier is analogous to compressing a solid. Eventually, the surface pressure falls away as the film collapses.

Dipping of a substrate and deposition of Langmuir–Blodgett films is carried out at a pressure within this 'solid' phase. Assuming the substrate has a hydrophilic surface, deposition will not occur when the substrate is dipped through the surface for the first time. However, a layer will deposit when the film is withdrawn (figure 9.10). The next layer will deposit on the second downwards movement and the orientation of the molecules will be reversed compared with that of the first layer. Stearic acid tails from one layer meet up with the tails from the other layer. Repeated dipping and withdrawal produces repeated double layers until the required overall thickness is achieved. It is interesting to compare the repeated double-layer structure as shown in figure 9.10 with the structure for the common black soap film shown in figure 5.5(b).

The potential of the method is that organic molecules can be deposited in a series of highly regular layers, provided the molecules do not rearrange themselves on the substrate.

It is possible to use a double-compartment trough with a sealing barrier between the two such that the substrate can be immersed in one compartment, passed underneath the barrier (with a supporting rod squeezing through the barrier) and withdrawn from the other compartment. With two different organic molecules spread across the surfaces within the two compartments, thin films of alternate layers can be produced. For a detailed account of Langmuir–Blodgett films see for instance Gaines (1966) or Roberts (1990).

9.8 APPLICATION IN TECHNOLOGY; COMMUNICATIONS NETWORKS

The property of a soap film in minimizing its length when confined between parallel plates can be used to design communications systems of minimum length. It is, of course, necessary to set the film into the correct pattern for the global energy minimum rather than merely a local minimum. For instance we can insert pins between the plates to represent the positions of towns or companies which we might wish to connect. We position the pins in scale with the actual geographic positions. Dipping the parallel plates into soap solution and withdrawing them (with perhaps a gentle blow on the resulting film) establishes the required pattern for connections. Such a process is perhaps not very applicable for a road system, where motorists will wish to keep their journey times low and will encourage the authorities to build a more complete system of interconnections. However, the approach can be more applicable to a network of pipelines or to an optical fibre telephone link where capital costs may be dominant. If holes are cut from one or both plates, the film cannot cross these regions. Hence, cutting holes from the plates can be used to represent impassable mountain ranges or large lakes that necessitate diversion of the communications system. However, we saw on page 29 that mathematical approaches to the problem produce approximate solutions to within 13% of the best possible solution, even when the number of points to be joined is large.

9.9 ARCHITECTURAL APPLICATIONS

Minimization of surface area can produce surfaces which are aesthetically satisfying and which can be reconstructed on a large scale because of the uniform distribution of forces. Some tent

water

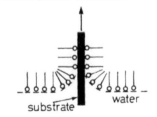
first withdrawal of substrate:

substrate   water

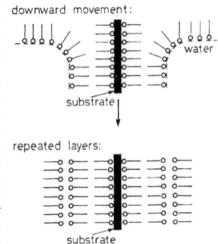
downward movement:

water

substrate

repeated layers:

substrate

Figure 9.10. Deposition of Langmuir–Blodgett layers by dipping a substrate.

structures can take on a similar appearance to soap-film minimal surfaces.

A number of architects have actively used ideas taken from minimal surface shapes. For instance, the German Pavilion at the 1967 World Exposition in Montreal had a roof structure consisting of a transparent pre-stressed membrane which was suspended from a number of mast-heads, supported by a steel-wire net, and anchored along a series of ridges and by a series of fixing points. The analogy with a soap film supported by a wire framework is excellent. The forces within the roof are tensional and there are compressive forces in the supporting poles and anchor points.

The Montreal Pavilion was designed by Frei Otto and Rolf Gutbrod. Other similar designs from this architectural practice have included the 1972 Olympic main stadium at Munich and also the Olympic swimming area and the athletics arena. These structures were designed to be easily erected and also easily dismantled and transported.

The architectural partnership actually used soap-film models to develop the designs. Vertical rods of differing heights were fixed into a perspex base-plate and these rods were joined by a series of threads. The whole model could be dipped into soap solution and the resulting minimal surface observed as the threads pulled taut. Solid models were then reproduced and fully tested, including testing them in a wind tunnel.

We saw in Chapter 5 how when we produced a catenoidal film between two hoops, we could not separate the hoops beyond a certain point and retain the catenoidal shape. An analogous problem is to see how high a soap film can be lifted from a single hoop when the second hoop has been replaced by a loop of thread (figure 9.11(a)). Such a model has direct application in designing high roof structures where the roof itself is in tension. For instance, if a roof structure is suspended in this way, it may be necessary to involve a number of loop supports (figure 9.11(b)).

9.10 CONCLUDING COMMENTS

One attractive feature of soap films is that they can be very easily studied in the laboratory, lecture room, or even in the kitchen. They form very complex, very beautiful, and aesthetically very pleasing shapes. What we see has come about from interactions on a molecular scale. Yet we have looked at and studied the

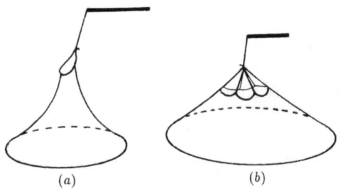

Figure 9.11 (a) Soap film suspended from a thread loop. (b) Soap film suspended from a number of thread loops.

patterning of the films without needing to know precise details about the interactions between the molecules.

Although the soap films are of such a size that we can hold them in our hands using a suitable framework, we can use them to help explain the formation and changes of crystal structures and of single biological cells on a microscopic scale. Moving to the scale of everyday life, they can suggest features in the design of communications networks and architectural roofs. On a macroscopic scale, we can draw analogy between soap films and the bending of light in space. Experiments with soap films are great fun and show how the underlying physics can explain both everyday and scientifically fundamental phenomena.

Appendix I

Construction of the models

A.I.1 THE TWO-DIMENSIONAL MODELS AND THE WEDGE

Most of the models described in this book were constructed in the workshop of the Physics Department at Essex University to withstand heavy usage over many years. However, the majority of the models can be made much more simply using inexpensive components. In most examples, the soap-film patterns do not depend on high precision construction of the plates or frameworks.

The two-dimensional models are best made from perspex or plexiglas. Various features of the construction are shown in figure A.I.1. Sheets of thickness 4 mm or 5 mm are convenient. In models used by the author for illustrating static two-dimensional patterns, the perspex sheets have been of size 18 cm by 18 cm with an inner separation of 12 mm. The fixed pins between the sheets can be metal rods (preferably of brass or stainless steel to avoid rusting) tightly fitting into pre-drilled holes. Even if the pins start out with a tight fit, there is a tendency for the plates to become loose and slip on the pins. The plates then cease to remain parallel. Plastic spacers can prevent this, but the effective pin size is increased by the presence of the surrounding plastic making the establishment of the correct film pattern more difficult. A better method is to use 'self-loc' pins. These are 'as-purchased' and consist of hollow cylindrical pins with a slot down the length. A size of $\frac{3}{32}$ inch (2.4 mm) diameter has been used. These hold tightly in pre-drilled holes cut very slightly smaller than the size of the expanded pins so that the pins are squeezed in. The pins need to be approximately $\frac{7}{8}$ inch (22 mm) long and can be cut to length.

An effective but more temporary method is to use stiff acetate sheet whilst the pins can be matchsticks. Acetate sheet as used for overhead projectors is suitable provided it is heavy gauge material. If very flexible sheet is used, the models distort and it is difficult to establish correct film patterns. The holes in the

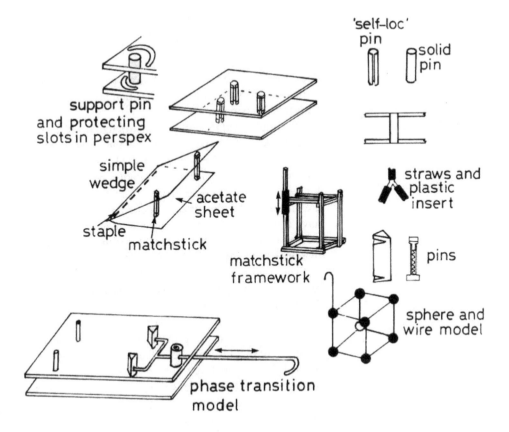

Figure A.I.1 Various constructional features of the models.

acetate sheet need to be inserted carefully (by using a needle or similar implement) to be of a size such that the matchsticks fit tightly. It is important to take care that the holes are aligned in each sheet by overlaying the sheets when producing the holes. It is also necessary to ensure that the sheets are sitting parallel after inserting the matchsticks.

The wedge can be constructed similarly from 4 mm perspex sheets of 24 cm by 15 cm. The apex of the wedge is held together with small stainless-steel or chromium-plated screws. Again the fixed pin can be 'self-loc'. The other pin moves along in grooves cut in the perspex sheets parallel to the apex. If the fixed pin and the grooves are positioned a distance of 12 cm out from the apex, the pin separation should alter from approximately 5 cm to 18 cm. The grooves must allow this. A strong handle needs to be soldered to the movable pin. 3 mm diameter brass or stainless steel is suitable. In the construction used by the author, a 12 mm

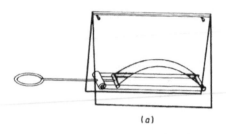

(a)

Figure 5.2. (a) *Revisited.* Wedge model with soap film.

diameter brass pillar is inserted into circular slots in the perspex and fixed by screws from the other side. The handle to the pin is able to slide back and forth in a hole inserted centrally in the pillar (see figures 5.2(a) and frontispiece F.1). The wedge angle does not affect the geometry but an angle of 20° has been used in practice corresponding to pillar heights of approximately 4 cm between the perspex sheets.

A simple wedge can be constructed from acetate sheet and matchsticks. The apex of the wedge is secured by a line of staples along one long edge of each of the sheets (which are typically 8 cm by 6 cm). Suitably positioned holes are cut for two matchsticks equidistant from the apex, the matchsticks are inserted and the wedge opened out. It is not practical to have a movable pin in this arrangement, but a series of models can be produced with a range of spacings between the pairs of matchsticks to illustrate the effect of changing the pin separation.

The two-dimensional models exhibiting phase transitions require movable pins. Various types have been used successfully. The more complicated type involves hollow pins attached to a wide foot. The pins contain a spring and a plastic insert which is, as a consequence, spring-loaded against the upper perspex plate. These pins can be either free-standing or attached to metal handles which are soldered to the pins. A rather simpler arrangement is to use triangular plastic pins. These tend to stand erect easily. Triangular pins of 8 mm side are successful. One end is cut on a slight incline and a saw cut inserted just below the highest point. This results in a flexible flat which produces the pressure fit against the upper perspex plate. Alternatively, if the pins are being attached to rods, then they can be left solid so that a push fit alone occurs between the parallel plates.

For the two-dimensional models illustrating phase changes, the plates can be set up with four separation pins at the corners to produce a totally versatile model. The pins need to be more substantial than the pins described previously. For instance a screw and nut arrangement can be used with the screws passing through perspex pillars or alternatively perspex columns can be cemented to the plates. These pins play no part in the patterns and any films linking up with these need to be broken with a drinking straw. Arcs cut from the perspex sheets to the sides of the pins help discourage erroneous film formation. For the first-order and second-order phase change models fixed pairs of 'self-loc' pins are set at 10 cm separation. 30 cm away, along the perpendicular bisector to the line joining the 'self-loc' pins, is located a 12 mm diameter pillar through which a long handle is inserted. To this is attached either a pair of pins

(10 cm separation) or a single pin to produce the first-order and second-order models respectively. Overall plate sizes to allow this arrangement should be 20 cm width and 35 cm length.

Similar pairs of parallel plates can be used for the optical refraction analogues. Two movable triangular pins are very suitable for establishing the soap film between two fixed points. The 'optical' inserts are also made from perspex, usually using perspex sheet of 4 mm thickness, 10 cm width, and say 28 cm length so that they are easy to hold. The actual cut-outs (consisting of a rectangle, a triangle (prism), or of the various types of lenses) can be approximately 8 cm long. The perspex inserts to model the refraction of P-waves and the bending of light near a black hole require the use of a lathe.

A.I.2 THE FRAMEWORKS

These have been constructed from a range of materials in a range of sizes, and each have their particular advantages. For instance, models have been constructed using 3 mm diameter brass or stainless steel (these produce large durable models for class demonstration) or using 1.5 mm diameter brass or stainless steel welding rod. Lengths of rod are soldered together. A mixture of hard-soldering of early joints followed by soft-soldering of later joints can aid construction. Bending of the rod to form some of the corners can produce more durable models although this may lead to less precision at the corners. It is important that excess solder does not remain at any corners as this prevents correct soap-film formation.

Fixed-dimension structures can be made alternatively from plastic straws and suitable plastic inserts for the corners†. Frameworks can also be made by using small spheres (as used for modelling crystal structures) with tiny holes cut in the spheres to allow insertion of thin stainless steel wire at the correct angles‡.

The variable frameworks need more careful construction. The base of each frame is made by bending brass or stainless steel rod in a plane to form the basal outline and soldering the two free ends together at the correct angle. Vertical sides are soldered at the corners of the base with one side extending further up than the others so that its end can eventually form a handle. Next is

† A kit for such models is distributed by Cochranes of Oxford Ltd, Leafield, Oxford OX8 5NT.
‡ Such models are supplied by Beevers Miniature Models, Chemistry Department, University of Edinburgh, Edinburgh EH9 3JJ.

required a short length of brass or stainless steel tubing (say 3 cm length) of internal diameter such that it can slide up and down this extended vertical side. A further length of rod is bent to form the outline of the top of the framework in one plane. The free ends of the rod are correctly soldered to the sliding tube such that the plane of the bent rods is perpendicular to the axis of the rod. The lengths of the sides of this top frame must be accurate, and the angles must be correct, so that when grooves are cut or filed into the outer edges of the corners, these corners can slide up and down against all the framework verticals with a precise fit. The rod is slipped onto the extended vertical such that the attached frame sits within the remainder of the framework. Finally the top end of the extended vertical (the vertical down which the tube slips) is bent both to form a handle and to prevent the two parts of the framework becoming separated. Figure F.2 (frontispiece) illustrates the details for the pentagonal prism. The stationary part of the framework is held by the handle and the sliding part is moved by sliding the length of tubing up and down the vertical side. Dimensions for models used by the author are approximately as follows: triangular edges for the triangular prism are 10 cm, square edges for the cuboid are 9 cm, and pentagonal edges for the pentagonal prism are 6 cm, whilst the verticals for each of the frameworks are approximately 10 cm.

Similar fixed frameworks can be constructed very simply by gluing together matchsticks. After construction, it is advisable to spray the frameworks with paint (using a waterproof aerosol paint). It is also possible to make variable models using matchsticks. The sliding component is a short length of plastic straw which slides up and down one of the matchstick sides. To this short length of straw is glued the upper part of the framework; otherwise construction is similar to the more elaborate metal models.

Helical frameworks and wire hoops can be constructed by suitably bending metal wire or rod. The saddle-point bifurcation model illustrated in Chapter 1 (page 11) can be constructed from 1.5 mm stainless steel welding rod to have a length of approximately 15 cm and widths in the other two directions of 3 cm. This achieves adequate overall rigidity plus enough flexibility to induce the switching of the saddle point.

As an alternative to using soap films to produce the minimal areas, small permanent models can be produced using quick-setting plastic film solution available in certain toy and model shops. Considerable trial-and-error is required to produce the optimum conditions.

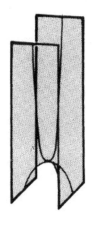

Figure 1.7. *Revisited.*

Appendix II

Soap-film patterns by computer using *Mathematica*

by John Tilley and David Lovett

The following programs all involve the graphics capability of *Mathematica*. The programs have been developed on Macintosh personal computers. Whereas *Mathematica* can be run on machines with 4 megabytes of RAM, the following programs normally require more RAM (say 9 megabytes) in order to carry out the graphical calculations.

Where multiple images are computed for animation, they have been set up within boxed outlines so that they are produced automatically to the same size. These are produced by the programs when **Evaluate Notebook** is highlighted under the **Action** menu. Before animation, the images should be highlighted and aligned on the left-hand side using **Align Selected Graphics**. The **Animation Settings** box is used to select the animation sequence. Where the effects discussed in this book show hysteresis, the programs do not produce a correct physical sequence if run in reverse. Loops should be set for repetition in a forward loop at say 2 frames per second. Animation will appear more continuous once the graphics have looped once.

The minimal surface programs are not animated but nevertheless require the larger amounts of RAM. They require the use of the program **Parametric plot 3D.m** available with the *Mathematica* packages. This program must be run first. Viewpoint setting are suggested for the minimal-surface plots. However, other settings can be selected. This is best achieved by selecting **3D ViewPoint Selector** in the **Prepare Input** submenu of the **Action** menu. Highlight the viewpoint settings in the program including the enclosing brackets. Move to the **3D Viewpoint Selector**. Now change the settings by changing the orientation of the cube. Once the desired viewpoint has been found, click the **Paste** button to insert the viewpoint figures.

It is advisable to clear the memory before running successive programs.

A.II.1–A.II.16 COMPUTER PROGRAMS

A.II.1 Variation of length (energy) of a soap film with configuration parameter.

A.II.1a Variation of length (energy) of a soap film with configuration parameter z; shows non-equilibrium film movement and point on energy curve.

A.II.2 Four-pin soap-film model of first-order phase change, showing animation of the pin movement and variation of length L with film configuration parameter x.

A.II.3 Four-pin soap-film model of first-order phase change, showing animation of the pin movement and variation of dL/dx with film configuration parameter x.

A.II.4 Three-pin soap-film model of second-order phase change, showing animation of the pin movement and variation of length L with film configuration parameter x.

A.II.5 Three-pin soap-film model of second-order phase change, showing animation of the pin movement and variation of gradient dL/dx with film configuration parameter x.

A.II.6 Three-pin soap-film model of second-order phase change, showing animation of the pin movement and variation of d^2L/dx^2 with film configuration x.

A.II.7 Three-pin soap-film model of second-order phase change, showing animation of the pin movement and variation of order parameter η with film configuration parameter x.

A.II.8 Four-pin soap-film phase transition and catastrophe model, showing animation of control-pin movement and variation of length L with control-pin position x.

A.II.9 Four-pin soap-film phase transition and catastrophe model, showing animation of control-pin movement and variation of gradient dL/dx with control-pin position x.

A.II.10 Four-pin soap-film phase transition and catastrophe model, showing animation of control-pin movement and variation of length L with angle θ as pin circulates around cusp point Z.

A.II.11 Four-pin soap-film phase transition and catastrophe model, showing animation of the control-pin movement and variation of $dL/d\theta$ with angle θ as control pin circulates around cusp point Z.

A.II.12 Helicoidal (spiral staircase) minimal surface.

A.II.13 Catenoidal minimal surface.

A.II.14 Scherk's minimal surface.

A.II.15 Catalan's minimal surface.

A.II.16 (i) Enneper's minimal surface. (ii) Enneper's minimal surface—alternative derivation using spherical coordinates.

(* A.II.1 VARIATION OF LENGTH (ENERGY) OF A SOAP FILM WITH CONFIGURATION PARAMETER*)

```
(* Four pins (two move together):
Plots Energy (film length L) vs. angle parameter z
for fixed pin separations.
Vertical pin separation = 2.0
Horizontal pin separation from 0.8 to 3.2 *)

SetOptions[Graphics,PlotRange->Automatic];

Do[

  xp=0.8 + i 0.4;
  leg=ToString[StringForm["separation = ``",xp]];

  phib=ArcTan[2./xp]/Degree//N;
  anga=N[(g[z] + phib) Degree,6];

  m1=Solve[{y-1==(x-xp) Tan[anga],y==0},{x}];
  xa=x/.m1[[1]];

  m2=Solve[{y-1==(x-xp) Tan[anga],x==xp/2.},{y}];
  ya=y/.m2[[1]];

  f1=N[4*Sqrt[(xp-xa)^2 + 1] + 2 xa - xp,6];
  f2=N[4*Sqrt[(xp/2.)^2 + (1-ya)^2] + 2 ya,6];

  zmax=N[Sqrt[85.- phib]];
  zmin=N[-Sqrt[phib - 5.]];

  g[z_]:= Which[z<0,-(z^2.),z>=0,z^2.];
  f3[z_]:=Which[z<0,f2,z>=0,f1];
```

```
  f3e=Table[f3[z],{z,zmin,zmax,zmax-zmin}];

  f3m=Max[f3e[[1]],f3e[[2]]];

  lpos=Which[xp<1.6,zmin+1.,xp>=1.6,1.0];

  ezplot:=Plot[f3[z],{z,zmin,zmax},
    DefaultColor->Hue[0.18],
    Background->Hue[0.66],
    DisplayFunction->Identity];

  Show[ezplot,Graphics[
    Text[leg,{lpos,f3m},{-1,0}]],
    DisplayFunction->$DisplayFunction],

  {i,0,6,1}]
```

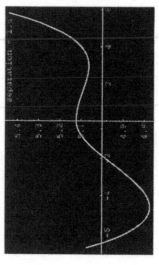

(* A.II.1a VARIATION OF LENGTH (ENERGY) OF A SOAP FILM WITH CONFIGURATION PARAMETER z SHOWS NON-EQUILIBRIUM FILM MOVEMENT AND CORRESPONDING MOVING POINT ON ENERGY CURVE*)

```
(* Four fixed pins:Draws varying non-equilibrium
film pattern.
Plots Energy (film length L) vs. angle parameter z.
Vertical pin separation = 2.0;
Horizontal pin separation = 3.0.
Other values of horizontal separation may be used, but
these may require new definitions of zmin & zmax *)

SetOptions[Graphics,Axes->None,
AspectRatio->Automatic,
Prolog->AbsolutePointSize[6],PlotRange->All,
PlotRegion->{{0,1},{0,1}}];

rt = N[Sqrt[3],6];
rtr = N[1/Sqrt[3],6];

xp = 3.0;
phib=ArcTan[2./xp]/Degree//N;
zmax=N[Sqrt[85.- phib]];
zmin=N[-Sqrt[phib - 5.]];

gf1 = Graphics[Line[{{-0.2,1.2},{xp+0.2,1.2},
{xp+0.2,-1.2},{-0.2,-1.2},{-0.2,1.2}}]];

q[z_]:= Which[z<0,-(z^2.),z>=0,z^2.];

anga:= N[(q[z] + phib) Degree,6];
m1=Solve[{y-1==(x-xp) Tan[anga],y==0},{x}];
xa=x/.m1[[1]];
m2=Solve[{y-1==(x-xp) Tan[anga],x==xp/2.},{y}];
ya=y/.m2[[1]];
list2 = {{Hue[0.4,1,0.5],Point[{xp,1}],Point[{xp,-1}],
    Point[{0,-1}],Point[{0,1}]},
    {Hue[0.3,1,1],Line[{{xp,1},{xa,0},{xp-xa,0},
    {0,1}}], Line[{{xp,-1},{xa,0}}]},
    {0,1}}], Line[{{xp,-1},{xp-xa,0}}]}};
g2 = Graphics[list2];

list3 = {{Hue[0.4,1,0.5],Point[{xp,1}],Point[{xp,-1}],
    Point[{0,-1}],Point[{0,1}]},
    {Hue[0.3,1,1],Line[{{xp,1},{xp/2.,ya},
    {xp/2.,-ya},{xp,-1}}],
    Line[{{0,1},{xp/2.,ya}}],
    Line[{{0,-1},{xp/2.,-ya}}]}};
g3 = Graphics[list3];
```

```
f1:=N[4*Sqrt[(xp-xa)^2 + 1] + 2 xa - xp,6];
f2:=N[4*Sqrt[(xp/2.)^2 + (1-ya)^2] + 2 ya,6];

f3[z_]:=Which[z<0,f2,z>=0,f1];

ezplot:=Plot[f3[z],{z,zmin,zmax},Axes->True,
    DefaultColor->Hue[0.18],
    Background->Hue[0.66],
    DisplayFunction->Identity];

mvpoint1:= Show[ezplot,Graphics[{Hue[0.2],
    PointSize[0.04],
    Point[{z,f1}]}]];

mvpoint2:= Show[ezplot,Graphics[{Hue[0.2],
    PointSize[0.04],
    Point[{z,f3[z]}]}]];

Do[
```

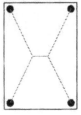

```
    Show[GraphicsArray[{Show[gf1,g3,
    DisplayFunction->Identity],mvpoint2}],
    DisplayFunction->$DisplayFunction],
    {z,zmin,0,-zmin/10}
]
```

```
Do[
    Show[GraphicsArray[{Show[gf1,g2,
    DisplayFunction->Identity],mvpoint1}],
    DisplayFunction->$DisplayFunction],
    {z,-zmin/10,zmax,-zmin/10}
]
```

```
(* A.II.2 FOUR-PIN SOAP-FILM MODEL OF FIRST-ORDER PHASE CHANGE, SHOWING
ANIMATION OF THE PIN MOVEMENT AND VARIATION OF FILM LENGTH L WITH FILM
CONFIGURATION PARAMETER x *)

(* Four Pins at the corners of a rectangle:
two fixed; two move together.
Draws film pattern, energy plot and point.
Produces animation. *)

SetOptions[Graphics,Axes->None,
AspectRatio->Automatic,
Prolog->AbsolutePointSize[6],PlotRange->All,
PlotRegion->{{0,1},{0,1}}];

rt = N[Sqrt[3],6];
rtr = N[1/Sqrt[3],6];

step = N[(4 - 2 rtr)/20];

gf1 = Graphics[Line[{{-0.2,1.2},{4.2,1.2},
{4.2,-1.2},{-0.2,-1.2},{-0.2,1.2}}]];

list2 = {{Hue[0.7,1,0.75],Point[{xm,1}],Point[{xm,-1}],
    Point[{0,-1}],Point[{0,1}]},
    {Hue[0.64],Line[{{xm,1},{xm-rtr,0},{rtr,0},
    {0,1}}],
    Line[{{xm,-1},{xm-rtr,0}}],Line[{{0,-1},
    {rtr,0}}]}};
g2 = Graphics[list2];

list3 = {{Hue[0.7,1,0.75],Point[{xm,1}],Point[{xm,-1}],
    Point[{0,-1}],Point[{0,1}]},
    {Hue[0.64],Line[{{xm,1},{xm/2,1.-(xm rtr/2)},
    {xm/2.,(xm rtr/2)-1.},{xm,-1}}],
    Line[{{0,1},{xm/2,1.-(xm rtr/2)}}],
    Line[{{0,-1},{xm/2,(xm rtr/2)-1.}}]}};
g3 = Graphics[list3];

f1[xp_]:= N[6 rtr + xp,6];
f2[xp_]:= N[rt xp + 2.,6];

f4[xp_]:= Which[xp<2 rtr,f2[xp],xp>=2 rtr,f1[xp]];
f5[xp_]:= Which[xp<2 rtr,f2[xp],xp>=2 rtr,f1[xp]];
```

```
eplot:= Plot[{f4[xp],f5[xp]},{xp,0,4.5},
PlotStyle->{Hue[0.4,0.9,0.75]},Axes->True,
DisplayFunction->Identity];

mvpoint1:= Show[eplot,Graphics[{PointSize[0.06],
Hue[0,1,1],Point[{xm,f1[xm]}]}]];

mvpoint2:= Show[eplot,Graphics[{PointSize[0.06],
Hue[0,1,1],Point[{xm,f2[xm]}]}]];

Do[

Show[GraphicsArray[{Show[gf1,g2,
DisplayFunction->Identity],mvpoint1],
DisplayFunction->$DisplayFunction],
{xm,4,2 rtr,-step}]
```

```
Do[

Show[GraphicsArray[{Show[gf1,g3,
DisplayFunction->Identity],mvpoint2],
DisplayFunction->$DisplayFunction],
{xm,2 rtr,2 rtr - 5 step,-step}]

Do[

Show[GraphicsArray[{Show[gf1,g3,
DisplayFunction->Identity],mvpoint2],
DisplayFunction->$DisplayFunction],
{xm,2 rt - 21 step,2 rt,step}]

Do[

Show[GraphicsArray[{Show[gf1,g2,
DisplayFunction->Identity],mvpoint1],
DisplayFunction->$DisplayFunction],
{xm,2 rt,2 rt + 4 step,step}]
```

(* A.II.3 FOUR-PIN SOAP-FILM MODEL OF FIRST-ORDER PHASE CHANGE, SHOWING ANIMATION OF THE PIN MOVEMENT AND VARIATION OF GRADIENT *dL/dx* WITH FILM CONFIGURATION PARAMETER *x* *)

```
(* Four Pins:
Draws film pattern, energy gradient
(dL/dx vs x) plot and point.
Produces animation *)

SetOptions[Graphics,Axes->None,
AspectRatio->Automatic,
Prolog->AbsolutePointSize[6],PlotRange->All,
PlotRegion->{{0,1},{0,1}}];
rt = N[Sqrt[3],6];
rtr= N[1/Sqrt[3],6];
step = N[(4 - 2 rtr)/20];
gf1=Graphics[Line[{{-0.2,1.2},{4.2,1.2},
   {4.2,-1.2},{-0.2,-1.2},{-0.2,1.2}}]];
list2 = {{Hue[0.7,1,0.75],Point[{xm,1}],
   Point[{xm,-1}],Point[{0,-1}],
   Point[{0,1}]},
   {Hue[0.64],Line[{{xm,1},{xm-rtr,0},
   {rtr,0},{0,1}}]],
   Line[{{xm,-1},{xm-rtr,0}}],
   Line[{{0,-1},{rtr,0}}]}};
g2 = Graphics[list2]
list3 = {{Hue[0.7,1,0.75],Point[{xm,1}],
   Point[{xm,-1}],Point[{0,-1}],
   Point[{0,1}]},{Hue[0.64],
   Line[{{xm,1},{xm/2.,1-(xm rtr/2)},
   {xm/2.,(xm rtr/2)-1},{xm,-1}}],
   Line[{{0,1},{xm/2.,1-(xm rtr/2)}}],
   Line[{{0,-1},{xm/2,(xm rtr/2)-1.}}]}};
g3 = Graphics[list3]
f4[xp_]:=Which[xp<=2 rtr,rt,xp>2 rtr,1];
f5[xp_]:=Which[xp<2    rt,rt,xp>=2 rt,1];
eplot:=Plot[{f4[xp],f5[xp]},{xp,0,4.5},
   Axes->True,AxesOrigin->{0,0.9},
   PlotRange->{{0,4.5},{0.9,1.8}},
   Ticks->{{{1.1,{0.02,0}},{2.2,{0.02,0}},
   {3.3,{0.02,0}},{4,4.{0.02,0}}},
   {{1.0,"1.0",{0.02,0}},{1.3,1.3,{0.02,0}},
   {1.6,1.6,{0.02,0}}}},
```

```
PlotStyle->{Dashing[{0.05,0.05}]},
   DefaultColor->Hue[0.18],
   Background->Hue[0.66],
   DisplayFunction->Identity];

mvpoint1:=Show[eplot,Graphics[{Hue[0.2],
   PointSize[0.06],Point[{xm,1}]}]];

mvpoint2:=Show[eplot,Graphics[{Hue[0.2],
   PointSize[0.06],Point[{xm,rt}]}]];
Do[
```

```
Show[GraphicsArray[{Show[gf1,g2,
   DisplayFunction->Identity],mvpoint1],
   DisplayFunction->$DisplayFunction},
   {xm,4,2 rtr,-step}]
```

```
Do[
Show[GraphicsArray[{Show[gf1,g3,
   DisplayFunction->Identity],mvpoint2],
   DisplayFunction->$DisplayFunction},
   {xm,2 rtr,2 rtr - 5 step,-step}]
```

```
Do[
Show[GraphicsArray[{Show[gf1,g3,
   DisplayFunction->Identity],mvpoint2],
   DisplayFunction->$DisplayFunction},
   {xm,2 rt - 21 step,2 rt,step}]
```

```
Do[
Show[GraphicsArray[{Show[gf1,g2,
   DisplayFunction->Identity],mvpoint1],
   DisplayFunction->$DisplayFunction},
   {xm,2 rt,2 rt + 4 step,step}]
```

```
(* A.II.4  THREE-PIN SOAP-FILM MODEL OF SECOND-ORDER PHASE CHANGE,
SHOWING ANIMATION OF THE PIN MOVEMENT AND VARIATION OF LENGTH L
WITH FILM CONFIGURATION PARAMETER x *)

(* Three pins, one moving.
Draws film pattern, energy plot and point.
Produces animation. *)

SetOptions[Graphics,Axes->None,
AspectRatio->Automatic,
Prolog->AbsolutePointSize[6],PlotRange->All,
PlotRegion->{{0,1},{0,1}}];

rt = N[Sqrt[3],6];
rrt = N[1/Sqrt[3],6];

ya=-0.6;

(* Other values of ya to try: 0, +0.3, -0.9. *)

m1=Solve[{(x+rrt)^2 + y^2==4./3.,y==ya},{x}];
xa=Max[x/.m1[[1]],x/.m1[[2]]];

g1=Graphics[{Line[{{-0.2,1.2},{2.5,1.2},
{2.5,-1.2},{-0.2,-1.2},{-0.2,1.2}}],
{GrayLevel[0.5],Circle[{-rrt,0},2 rrt,
{-Pi/3,Pi/3}]},
{Dashing[{0.05,0.05}],Line[{{0,ya},
{2.4,ya}}]}}];

m2= Solve[{(x+rrt)^2 + y^2==4./3.,
        y(xm + rt)==ya(x +rt)},{x,y}];
xj= Max[x/.m2[[1,1]],x/.m2[[2,1]]];
yj= If[ya>=0.,Max[y/.m2[[1,2]],y/.m2[[2,2]]],
       Min[y/.m2[[1,2]],y/.m2[[2,2]]]];

list2 = {{Hue[0.8,1,1],Point[{0,1}],Point[{0,-1}],
          Point[{xm,ya}]},
         {Hue[0.8,0.7,1],Line[{{0,1},{xm,ya},
          {0,-1}}]}};
g2 = Graphics[list2];

list3 = {{Hue[0.8,1,1],Point[{0,1}],Point[{0,-1}],
          Point[{xm,ya}]},
         {Hue[0.8,0.7,1],Line[{{0,1},{xj,yj},
          {xm,ya}]},
          Line[{{0,-1},{xj,yj}}]}};
g3 = Graphics[list3];
```

```
f1[xp_]:=N[Sqrt[xp^2 +(1-ya)^2] +
                 Sqrt[xp^2 + (1+ya)^2],6];

f2[xp_]:=N[Sqrt[(xp+rt)^2 + ya^2],6];

f3[xp_]:=Which[xp<=xa,f1[xp],xp>xa,f2[xp]];

eplot=Plot[f3[xp],{xp,0,1.45},
PlotStyle->{Hue[0.4,1,0.75]},Axes->True,
AxesOrigin->{0,1.8},PlotRange->{{0,1.45},
{1.8,3.2}},
Ticks->{{{0.5,0.5,{0.02,0)},{1,"1.0",
{0.02,0}},{{2.0,"2.0",{0.02,0}},
{2.5,2.5,{0.02,0}},{3.0,"3.0",{0.02,0}}}},
DisplayFunction->Identity];

mvpoint1=Show[eplot,Graphics[{PointSize[0.05],
Hue[0,1,1],Point[{xm,f3[xm]}]}]];

mvpoint2=Show[eplot,Graphics[{PointSize[0.05],
Hue[0,1,1],Point[{xm,f3[xm]}]}]];
```

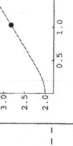

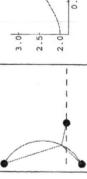

```
Do[

Show[GraphicsArray[{Show[g1,g3,
DisplayFunction->Identity],mvpoint1],
DisplayFunction->$DisplayFunction],
{xm,1.4,xa,(xa - 1.4)/8}]

Do[

Show[GraphicsArray[{Show[g1,g2,
DisplayFunction->Identity],mvpoint1],
DisplayFunction->$DisplayFunction],
{xm,xa,0.1,(0.1 - xa)/3}]
```

(* **A.II.5 THREE-PIN SOAP-FILM MODEL OF SECOND-ORDER PHASE CHANGE, SHOWING ANIMATION OF THE PIN MOVEMENT AND VARIATION OF GRADIENT** *dL/dx* **WITH FILM CONFIGURATION PARAMETER** *x* *)

```
(* Three pins, one moving.
Draws film pattern, energy gradient plot and point.
Produces animation. *)

(* Copy program A.II.4 as far as
       g3=Graphics[list3]; *)

f1[xp_]:=Sqrt[xp^2 +(1-ya)^2] +
          Sqrt[xp^2 + (1+ya)^2];

f2[xp_]:=Sqrt[(xp+it)^2 + ya^2]

f6[xp_]:=Which[xp<=xa,f1'[xp],xp>xa,f2'[xp]];

eplot:=Plot[f6[xp],{xp,0,1.45},
    PlotStyle->{Hue[0.4,1.0,0.75]},Axes->True,
    AxesOrigin->{0,0.3},PlotRange->{{0,1.45},
    {0.3,1.0}},
    Ticks->{{{0.5,0.5,{0.02,0}},
    {1,"1.0",{0.02,0}}},{{0.4,0.4,{0.02,0}},
    {0.6,0.6,{0.02,0}},{0.8,0.8,{0.02,0}},
    {1.0,"1.0",{0.02,0}}}},
    DisplayFunction->Identity];

mvpoint1:=Show[eplot,Graphics[{PointSize[0.05],
    Hue[0,1,1],Point[{xm,f6[xm]}]}]];

step=N[(xa - 1.4)/8];
```

```
Do[
Show[GraphicsArray[{{Show[g1,g3,
  DisplayFunction->Identity],mvpoint1}],
  DisplayFunction->$DisplayFunction],
  {xm,1.4,xa-step,step}]
```

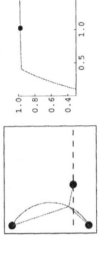
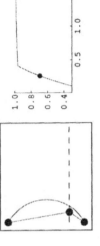

```
Do[
Show[GraphicsArray[{{Show[g1,g2,
  DisplayFunction->Identity],mvpoint1}],
  DisplayFunction->$DisplayFunction],
  {xm,xa,0.1,(0.1 - xa)/6}]
```

(* A.II.6 THREE-PIN SOAP-FILM MODEL OF SECOND-ORDER PHASE CHANGE, SHOWING ANIMATION OF THE PIN MOVEMENT AND VARIATION OF d^2L/dx^2 WITH FILM CONFIGURATION PARAMETER x *)

```
(* Three pins, one moving.
Draws film pattern, plot of second differential of
energy, and point.
Produces frames for animation. *)

(* Copy program A.II.4 as far as
   g3=Graphics[list3]; *)

f1[xp_]:=Sqrt[xp^2 +(1-ya)^2] +
          Sqrt[xp^2 + (1+ya)^2];

f2[xp_]:=Sqrt[ (xp+rt)^2 + ya^2];

f6[xp_]:=Which[xp<xa,f1''[xp],xp>=xa,f2''[xp]];

eplot:=Plot[f6[xp],{xp,0,1.45},
       PlotStyle->{Hue[0.4,1,0.75]},Axes->True,
       AxesOrigin->{0,-0.2},PlotRange->{{0,1.45},
       {-0.2,3.3}},
       Ticks->{{{-0.5,0.5,{0.02,0}},
       {1,"1.0",{0.02,0}}},{{0.0,"0.0",{0.02,0}},
       {1.0,"1.0",{0.02,0}},{2.0,"2.0",{0.02,0}},
       {3.0,"3.0",{0.02,0}}}},
       DisplayFunction->Identity];

mvpoint1:=Show[eplot,Graphics[{PointSize[0.05],
       Hue[0,1,1],Point[{xm,f6[xm]}]}]];
```

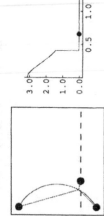

```
Do[
  show[GraphicsArray[{Show[g1,g3],
  DisplayFunction->Identity],mvpoint1]},
  DisplayFunction->$DisplayFunction],
  {xm,1.4,xa,(xa - 1.4)/8}]

Do[
  Show[GraphicsArray[{Show[g1,g2,
  DisplayFunction->Identity],mvpoint1]},
  DisplayFunction->$DisplayFunction],
  {xm,xa-0.01,0.09,(0.1 - xa)/4}]
```

(* A.II.7 THREE-PIN SOAP-FILM MODEL OF SECOND-ORDER PHASE CHANGE,
SHOWING ANIMATION OF THE PIN MOVEMENT AND VARIATION OF *ORDER
PARAMETER* WITH FILM CONFIGURATION PARAMETER *x* *)

```
(* Three pins, one moving.
Draws film pattern, plot of order parameter,
and point.
Produces frames for animation. *)

SetOptions[Graphics,Axes->None,
AspectRatio->Automatic,
Prolog->AbsolutePointSize[6],PlotRange->All,
PlotRegion->{{0,1},{0,1}}];

rt = N[Sqrt[3],6];
rrt = N[1/Sqrt[3],6];

ya=0;

(* Other values of ya to try: 0.3, -0.6, -0.8. *)

(* Copy program  A.II.4
   from    m1=
   to      g3=Graphics[list3]; *)

f1[xp_]:=N[Sqrt[ArcCos[(2 xp^2 +(1-ya)^2
        + (1+ya)^2 - 4)/
        (2*Sqrt[xp^2 +(1-ya)^2]*
        Sqrt[xp^2 + (1+ya)^2]]]*(3/Pi) - 2],6];

f3[xp_]:=Chop[Which[xp<=xa,f1[xp],xp>xa,0],10^-3];

eplot:=Plot[f3[xp],{xp,0,1.4},
    PlotStyle->{Hue[0.4,1.0,0.75]},
    Axes->True,AxesOrigin->{0,-0.1},
    PlotRange->{{0,1.45},{-0.1,1}},
    Ticks->{{{0.5,0.5,{0.02,0}},{1,"1.0",
    {0.02,0}},{{0.0,"0.0",{0.02,0}},
    {0.5,0.5,{0.02,0}},{0.5,0.5,{0.02,0}},
    {1.0,"1.0",{0.02,0}}}},
    DisplayFunction->Identity];

mvpoint1:=Show[eplot,Graphics[{PointSize[0.05],
    Hue[0,1,1],Point[{xm,f3[xm]}]}]];
```

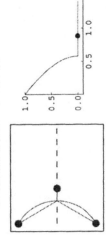

```
Do[
    Show[GraphicsArray[{{Show[g1,g3,
    DisplayFunction->Identity],mvpoint1},
    DisplayFunction->$DisplayFunction],
    {xm,1.4,xa,(xa - 1.4)/8}]
```

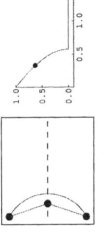

```
Do[
    Show[GraphicsArray[{{Show[g1,g2,
    DisplayFunction->Identity],mvpoint1},
    DisplayFunction->$DisplayFunction],
    {xm,xa-0.01,0.09,(0.1 - xa)/6}]
```

```
(* A.II.8  FOUR-PIN SOAP-FILM PHASE TRANSITION AND CATASTROPHE MODEL,
SHOWING ANIMATION OF CONTROL PIN MOVEMENT AND VARIATION OF LENGTH L
WITH CONTROL PIN POSITION x *)

(* Three pins fixed at the corners of an
equilateral triangle.
Fourth pin moves on a straight line close
to the circumcentre of triangle.
Draws equilibrium film pattern and calculates
film length.
Produces frames for animation.*)

SetOptions[Graphics,Axes->None,
AspectRatio->Automatic,
Prolog->AbsolutePointSize[6],PlotRange->All,
PlotRegion->{{0,1},{0,1}}];

rt = N[Sqrt[3],6];
rrt = N[Sqrt[1/3],6];

xc = 0.1;

g0 = Graphics[{Line[{{-0.65,0.55},{0.35,0.55},
     {0.35,-0.55},{-0.65,-0.55},{-0.65,0.55}}],
     {Dashing[{0.02,0.02}],Line[{{xc,0.2},
     {xc,-0.2}}]},
     {GrayLevel[0.5],Point[{0,0}]}}];

m3 = Solve[{(xc + rrt/2)^2 + (y + 0.5)^2
     == 1./3.,{y}];
yd = Max[y/.m3[[1]],y/.m3[[2]]];

m4 = Solve[{(xc + rrt/2)^2 + (y - 0.5)^2
     == 1./3.,{y}];
yu = Min[y/.m4[[1]],y/.m4[[2]]];

fm1[yc_] := N[Sqrt[(xc + rrt/2)^2 + (yc + 1)^2] +
     Sqrt[(xc - rrt/2)^2 + (yc - 0.5)^2],6];

fm2[yc_] := N[Sqrt[(xc + rrt/2)^2 + (yc - 1)^2] +
     Sqrt[(xc - rrt/2)^2 + (yc + 0.5)^2],6];

fm3[yc_] := Which[yc >= yd,fm1[yc],yc < yd,fm2[yc]];

fm4[yc_] := Which[yc >= yu,fm1[yc],yc < yu,fm2[yc]];
```

```
eplot := Plot[{fm3[yc],fm4[yc]},{yc,0.2,-0.2},
     Axes->True,AxesOrigin->{0,1.725},
     PlotRange->{{-0.2,0.2},{1.725,1.755}},
     Ticks->{{{-0.2,-0.2,{0.02,0}},
     {0.0,0},{0.2,0.2,{0.02,0}},
     {{1.730,"1.730",{0.02,0}},
     {1.755,1.755,{0.02,0}}}},
     PlotStyle->{Hue[0.7,1.0,0.75]},
     DisplayFunction->Identity];

mvpoint1 := Show[eplot,Graphics[{Hue[0.8,1,0.5],
     PointSize[0.06],Point[{ym,fm1[ym]}]}]];

mvpoint2 := Show[eplot,Graphics[{Hue[0.8,1,0.5],
     PointSize[0.06],Point[{ym,fm2[ym]}]}]];

Do[
  m1 = Solve[{(x + rrt/2)^2 + (y + 0.5)^2==1./3.,
       (y - ym)(xc + rrt) == (ym - xc)(x - xc)},
       {x,y}];
  x1j = Max[x/.m1[[1,1]],x/.m1[[2,1]]];
  y1j = Max[y/.m1[[1,2]],y/.m1[[2,2]]];

  list1 = {{Hue[0.7,1,0.75],Point[{-rrt,0}],
       Point[{xc,ym}],Point[{rrt/2,-0.5}]},
       Point[{rrt/2,0.5}]},
       {Hue[0.64],Line[{{-rrt,0},{x1j,y1j}},
       {rrt/2,-0.5}}],
       Line[{{x1j,y1j},{xc,ym},{rrt/2,0.5}}]}};
  g1 = Graphics[list1];

  Show[GraphicsArray[{Show[g0,g1,
       DisplayFunction->Identity],mvpoint1}],
       DisplayFunction->$DisplayFunction],

  {ym,0.2,yd,-(0.2 - yd)/8}]
```

(* A.II.8 FOUR-PIN SOAP-FILM PHASE TRANSITION AND CATASTROPHE MODEL,
SHOWING ANIMATION OF CONTROL PIN MOVEMENT AND VARIATION OF LENGTH *L*
WITH CONTROL PIN POSITION *x* * continued)

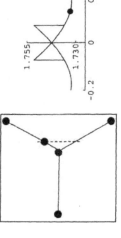

```
Do[

m2=Solve[{(x+rrt/2)^2 + (y-0.5)^2==1./3.,
         (y-ym)(xc+rrt)==(ym-1)(x-xc)},{x,y}];
x2j=Max[x/.m2[[1,1]],x/.m2[[2,1]]];
y2j=Min[y/.m2[[1,2]],y/.m2[[2,2]]];

list2 = {{Hue[0.7,1,0.75],Point[{-rrt,0}],
         Point[{xc,ym}],Point[{rrt/2,0.5}]},
         Point[{rrt/2,0.5}]},
         {Hue[0.64],Line[{{-rrt,0},{x2j,y2j},
         rrt/2,+0.5}]},
         Line[{{x2j,y2j},{xc,ym},{rrt/2,-0.5}}]}};
g2=Graphics[list2];

Show[GraphicsArray[{Show[g0,g2,
DisplayFunction->identity],mvpoint2}],
DisplayFunction->$DisplayFunction],
{ym,yd,-0.2,-(0.2 + yd)/5}]

Do[

m2=Solve[{(x+rrt/2)^2 + (y-0.5)^2==1./3.,
         (y-ym)(xc+rrt)==(ym-1)(x-xc)},{x,y}];
x2j=Max[x/.m2[[1,1]],x/.m2[[2,1]]];
y2j=Min[y/.m2[[1,2]],y/.m2[[2,2]]];
```

```
list2 = {{Hue[0.7,1,0.75],Point[{-rrt,0}]},
         Point[{xc,ym}],Point[{rrt/2,-0.5}],
         Point[{rrt/2,0.5}]},
         {Hue[0.64],Line[{{-rrt,0},{x2j,y2j},
         {rrt/2,+0.5}}]},
         Line[{{x2j,y2j},{xc,ym},{rrt/2,-0.5}}]}};
g2=Graphics[list2];

Show[GraphicsArray[{Show[g0,g2,
DisplayFunction->Identity],mvpoint2}],
DisplayFunction->$DisplayFunction],
{ym,-0.2,yu,(0.2-yd)/8}]

m1=Solve[{(x+rrt/2)^2 + (y+0.5)^2==1./3.,
         (y-ym)(xc+rrt) ==(ym+1)(x-xc)},{x,y}];
x1j=Max[x/.m1[[1,1]],x/.m1[[2,1]]];
y1j=Max[y/.m1[[1,2]],y/.m1[[2,2]]];

list1 = {{Hue[0.7,1,0.75],Point[{-rrt,0}],
         Point[{xc,ym}],Point[{rrt/2,-0.5}],
         Point[{rrt/2,0.5}]},
         {Hue[0.64],Line[{{-rrt,0},{x1j,y1j},
         {rrt/2,-0.5}}]},
         Line[{{x1j,y1j},{xc,ym},{rrt/2,0.5}}]}};
g1 = Graphics[list1];

Show[GraphicsArray[{Show[g0,g1,
DisplayFunction->Identity],mvpoint1}],
DisplayFunction->$DisplayFunction],
{ym,yu,0.2,(0.2 + yd)/5}]
```

(* A.II.9 FOUR-PIN SOAP-FILM PHASE TRANSITION AND CATASTROPHE MODEL, SHOWING ANIMATION OF CONTROL PIN MOVEMENT AND VARIATION OF GRADIENT dL/dx WITH CONTROL PIN POSITION x *)

```
(* Three pins fixed at the corners of an
equilateral triangle.
Fourth pin moves close to the circumcentre of
triangle.
Draws equilibrium film pattern, energy gradient
(dL/dx vs x) plot and point.
Produces frames for animation.*)

(*Copy program A.II.8 as far as
          yu=Min[y/.m4[[1]],y/.m4[[2]]];*)

fm1[yc_]=Sqrt[(xc+rrt)^2 + (yc+1)^2] +
          Sqrt[(rrt/2-xc)^2 + (yc-0.5)^2];

fm2[yc_]=Sqrt[(xc+rrt)^2 + (yc-1)^2] +
          Sqrt[(rrt/2-xc)^2 + (yc+0.5)^2];

gr3[yc_]:=Which[yc>=yd,fm1'[yc],yc<yd,fm2'[yc]];

gr4[yc_]:=Which[yc>=yu,fm1'[yc],yc<yu,fm2'[yc]];

eplot:=Plot[{gr3[yc],gr4[yc]},{yc,0.2,-0.2},
          Axes->True,PlotRange->Automatic,
          Ticks->{{-0.2,-0.2},{0.2,0.2}},
          {{-1.0,-1.0},{1.0,1.0}},
          PlotStyle->Hue[0.66],
          DisplayFunction->Identity];

mvpoint1:=Show[eplot,Graphics[{Hue[0],
          PointSize[0.06],Point[{ym,fm1'[ym]}]}]];

mvpoint2:=Show[eplot,Graphics[{Hue[0],
          PointSize[0.06],Point[{ym,fm2'[ym]}]}]];

Do[
    m1=Solve[{(x+rrt/2)^2 + (y+0.5)^2==1./3.,
          (y-ym)(xc+rrt) ==(ym+1)(x-xc)},{x,y}];
    x1j=Max[x/.m1[[1,1]],x/.m1[[2,1]]];
    y1j=Max[y/.m1[[1,2]],y/.m1[[2,2]]];
```

```
list1 = {{Hue[0.7,1,0.75],Point[{-rrt,0}],
          Point[{xc,ym}],Point[{rrt/2,-0.5}],
          Point[{rrt/2,0.5}]},
          {Hue[0.64],Line[{{-rrt,0},{x1j,y1j},
          {rrt/2,-0.5}}]},
          Line[{{x1j,y1j},{xc,ym},{rrt/2,0.5}}]}};
g1 = Graphics[list1];

Show[GraphicsArray[{Show[g0,g1,
          DisplayFunction->Identity],mvpoint1}],
          DisplayFunction->$DisplayFunction],
          {ym,0.2,yd,-(0.2 - yd)/8}]

(* Copy last three Do loops from A.II.8 *)
```

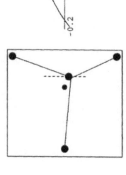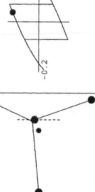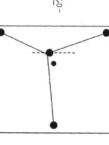

```
(* A.II.10  FOUR-PIN SOAP-FILM PHASE TRANSITION AND CATASTROPHE MODEL,
SHOWING ANIMATION OF CONTROL PIN MOVEMENT AND VARIATION OF LENGTH L
WITH ANGLE θ AS CONTROL PIN CIRCULATES ROUND CUSP POINT Z. *)
```

```
(* Three pins fixed at the corners of an
equilateral triangle.
Fourth pin moves close to the circumcentre of
triangle in a complete circle.
Draws equilibrium film pattern, length (energy)
as function of angle and point.
Produces frames for animation.*)

(* This program takes a long time to draw all
the frames and takes a significant amount of
memory - up to 5 MB.  Both time and memory use
can be reduced by increasing step size. *)

SetOptions[Graphics,Axes->None,
AspectRatio->Automatic,
Prolog->AbsolutePointSize[6],PlotRange->All,
PlotRegion->{{0,1},{0,1}}];

rt = N[Sqrt[3],6];
rrt = N[Sqrt[1/3],6];

cr= 0.1;

g0=Graphics[{Line[{{-0.65,0.55},{0.35,0.55},
{0.35,-0.55},{-0.65,-0.55},{-0.65,0.55}}],
{AbsolutePointSize[4],Point[{0,0}]},
Circle[{0,0},cr],{Hue[0.7,1,0.75],
Point[{-rrt,0}],Point[{rrt/2,0.5}],
{GrayLevel[0.5],
Circle[{-rrt/2,-0.5},rrt,{0,2*Pi/3//N}],
Circle[{-rrt/2,0.5},rrt,{-2*Pi/3/N,0}],
Circle[{rrt,0},rrt,{2*Pi/3//N,4*Pi/3//N}]}},
PlotRange->{{-0.65,0.35},{-0.55,0.55}}];

(* The following sections of the program find the
angles at which the small circle, around which the
fourth pin moves, intersects the large circular arcs
which join pairs of pins. It is on these arcs that
the equilibrium film junctions lie. The
intersection points represent the pin positions
at which the film configurations jump. These
positions are different for clockwise and
anticlockwise motion. *)
```

```
mc1=Solve[{((x+rrt/2)^2 + (y+0.5)^2==1./3.,
       x^2 + y^2 == cr^2},(x,y)];

xc1=Min[x/.mc1[[1]],x/.mc1[[2]]];
yc1=Max[y/.mc1[[1]],y/.mc1[[2]]];
ac1=ArcTan[xc1,yc1];

mc2=Solve[{((x+rrt/2)^2 + (y-0.5)^2==1./3.,
       x^2 + y^2 == cr^2},(x,y)];

xc2=Max[x/.mc2[[1]],x/.mc2[[2]]];
yc2=Max[y/.mc2[[1]],y/.mc2[[2]]];
ac2=ArcTan[xc2,yc2];

mc3=Solve[{((x-rrt)^2 + y^2==1./3.,
       x^2 + y^2 == cr^2},(x,y)];

xc3=Max[x/.mc3[[1]],x/.mc3[[2]]];
yc3=Min[y/.mc3[[1]],y/.mc3[[2]]];
ac3=ArcTan[xc3,yc3];

cw1=-ac2;
cw2=-ac1;
cw3=-ac3;
```

```
list1:= {{Hue[0.7,1,0.75],Point[{xc,yc}]},
{Hue[0.64],Line[{{-rrt,0},{x1j,y1j},
{rrt/2,-0.5}}]},
Line[{{x1j,y1j},{xc,yc},{rrt/2,0.5}}]};
g1:=Graphics[list1];

list2:= {{Hue[0.7,1,0.75],Point[{xc,yc}]},
{Hue[0.64],Line[{{-rrt,0},{x2j,y2j},
{rrt/2,+0.5}}]},
Line[{{x2j,y2j},{xc,yc},{rrt/2,-0.5}}]};
g2:=Graphics[list2];

list3:= {{Hue[0.7,1,0.75],Point[{xc,yc}]},
{Hue[0.64],Line[{{rrt/2,-0.5},{x3j,y3j},
{rrt/2,+0.5}}]},
Line[{{x3j,y3j},{xc,yc},{-rrt,0}}]};
g3:=Graphics[list3];
```

```
(* A.II.10  FOUR-PIN SOAP-FILM PHASE TRANSITION AND CATASTROPHE MODEL,
SHOWING ANIMATION OF CONTROL PIN MOVEMENT AND VARIATION OF LENGTH L
WITH ANGLE θ AS CONTROL PIN CIRCULATES ROUND CUSP POINT Z continued. *)


f11[aphi_]:=N[Sqrt[(cr Cos[aphi] + rrt)^2 +
    (cr Sin[aphi] + 1)^2] +
    Sqrt[(rrt/2 - cr Cos[aphi])^2 +
    (cr Sin[aphi] - 0.5)^2],6];

f12[aphi_]:=N[Sqrt[(cr Cos[aphi] + rrt)^2 +
    (cr Sin[aphi] - 1)^2] +
    Sqrt[(rrt/2 - cr Cos[aphi])^2 +
    (cr Sin[aphi] + 0.5)^2],6];

f13[aphi_]:=N[Sqrt[(cr Cos[aphi]-2 rrt)^2 +
    (cr Sin[aphi])^2] +
    Sqrt[(rrt+cr Cos[aphi])^2 +
    (cr Sin[aphi])^2],6];

f14[aphi_]:=Which[aphi>ac2 && aphi<=ac1,f11[aphi],
    aphi>ac1 || aphi<=ac3,f13[aphi],
    True,f12[aphi]];

f15[aphi_]:=Which[aphi>cw1 && aphi<=cw3,
    f11[aphi],aphi>cw3 || aphi<=cw2,
    f13[aphi],True,f12[aphi]];

eplot:=Plot[{f14[aphi],f15[aphi]},{aphi,-Pi,Pi},
    Axes->True,PlotStyle->{Hue[0.7,1,0.5]},
    Ticks->{{-Pi,{0,"0"},Pi},None},
    DisplayFunction->Identity];

mvpoint1:=Show[eplot,Graphics[{PointSize[0.04],
    Hue[0,1,1],Point[{zm,f14[zm]}]}]];

mvpoint2:=Show[eplot,Graphics[{PointSize[0.04],
    Hue[0,1,1],Point[{zm,f15[zm]}]}]];

Do[
    step=Pi/8//N;
    zm=-Pi + j*step//N;

    xc=cr Cos[zm];
    yc=cr Sin[zm];


    m1=Solve[{((x+rrt/2)^2 + (y+0.5)^2==1./3.,
        (y-yc)(xc+rrt) ==(yc+1)(x-xc)),{x,y}];
    x1j=Max[x/.m1[[1]],x/.m1[[2]]];
    y1j=Max[y/.m1[[1]],y/.m1[[2]]];

    m2=Solve[{((x+rrt/2)^2 + (y-0.5)^2==1./3.,
        (y-yc)(xc+rrt) ==(yc-1)(x-xc)),{x,y}];
    x2j=Max[x/.m2[[1]],x/.m2[[2]]];
    y2j=Min[y/.m2[[1]],y/.m2[[2]]];

    m3=Solve[{((x-rrt)^2 + y^2==1./3.,
        (y-yc)(xc-2 rrt)==yc (x-xc)),{x,y}];
    x3j=Min[x/.m3[[1]],x/.m3[[2]]];
    y3j=If[yc>=0.,Max[y/.m3[[1]],y/.m3[[2]]],
        Min[y/.m3[[1]],y/.m3[[2]]]];

    ga1:=Show[GraphicsArray[{Show[g0,g1,
        DisplayFunction->Identity],mvpoint1}],
        DisplayFunction->$DisplayFunction];

    ga2:=Show[GraphicsArray[{Show[g0,g2,
        DisplayFunction->Identity],mvpoint1}],
        DisplayFunction->$DisplayFunction];

    ga3:=Show[GraphicsArray[{Show[g0,g3,
        DisplayFunction->Identity],mvpoint1}],
        DisplayFunction->$DisplayFunction];

    If[zm<ac3 || zm>=ac1,ga3,
        If[zm>=ac3 && zm<ac2,ga2,ga1]],
    {j,0,16,1}]
```

(* A.II.10 FOUR-PIN SOAP-FILM PHASE TRANSITION AND CATASTROPHE MODEL, SHOWING ANIMATION OF CONTROL PIN MOVEMENT AND VARIATION OF LENGTH L WITH ANGLE θ AS CONTROL PIN CIRCULATES ROUND CUSP POINT Z continued. *)

```
ga2:=Show[GraphicsArray[{Show[g0,g2,
    DisplayFunction->Identity],mvpoint2}],
    DisplayFunction->$DisplayFunction];

ga3:=Show[GraphicsArray[{Show[g0,g3,
    DisplayFunction->Identity],mvpoint2}],
    DisplayFunction->$DisplayFunction];

If[zm<cw1 || zm>=cw3,ga3,
    If[zm]==cw2 && zm<cw3,ga1,ga2]],
    {j,0,16,1}]
```

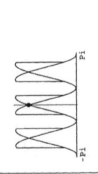

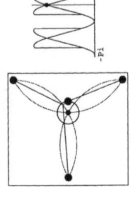

```
Do[

step=Pi/8//N;
zm=Pi - j*step//N;

xc=cr Cos[zm];
yc=cr Sin[zm];

m1=Solve[{((x+rrt/2)^2 + (y+0.5)^2==1./3.,
    (y-yc)(xc+rrt) ==(yc+1)(x-xc)},{x,y}];
x1j=Max[x/.m1[[1]],x/.m1[[2]]];
y1j=Max[y/.m1[[1]],y/.m1[[2]]];

m2=Solve[{((x+rrt/2)^2 + (y-0.5)^2==1./3.,
    (y-yc)(xc+rrt)==(yc-1)(x-xc)},{x,y}];
x2j=Max[x/.m2[[1]],x/.m2[[2]]];
y2j=Min[y/.m2[[1]],y/.m2[[2]]];

m3=Solve[{((x-rrt)^2 + y^2==1./3.,
    (y-yc)(xc-2 rrt)==yc (x-xc)},{x,y}];
x3j=Min[x/.m3[[1]],x/.m3[[2]]];
y3j=If[yc>=0.,Max[y/.m3[[1]],y/.m3[[2]]],
    Min[y/.m3[[1]],y/.m3[[2]]]];

ga1:=Show[GraphicsArray[{Show[g0,g1,
    DisplayFunction->Identity],mvpoint2}],
    DisplayFunction->$DisplayFunction];
```

(* A.II.11 FOUR-PIN SOAP-FILM PHASE TRANSITION AND CATASTROPHE MODEL, SHOWING ANIMATION OF CONTROL PIN MOVEMENT AND VARIATION OF GRADIENT $dL/d\theta$ WITH ANGLE θ AS CONTROL PIN CIRCULATES ROUND CUSP POINT Z. *)

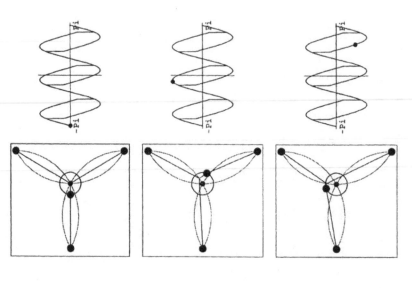

```
(* Three pins fixed at the corners of an
equilateral triangle.
Fourth pin moves close to the circumcentre of
triangle in a complete circle.
Draws equilibrium film pattern, gradient of length
(dL/dq) as function of angle, and point.
Produces frames for animation.*)

(* This program takes a long time to draw all
the frames and takes a significant amount of
memory - up to 5 MB. Both time and memory use
can be reduced by increasing step size. *)
(* Copy program A.II.10 as far as
    fl3[aphi_]:=N[Sqrt[(cr Cos[aphi]-2 rrt)^2 +
                (cr Sin[aphi])^2]
              + Sqrt[(rrt+cr Cos[aphi])^2
              +(cr Sin[aphi])^2],6];*)

gr4[aphi_]:=Which[aphi>ac2 && aphi<=ac1,fl1'[aphi],
                  aphi>ac1 || aphi<=ac3,fl3'[aphi],
                  True,fl2'[aphi]];

gr5[aphi_]:=Which[aphi>cw1 && aphi<=cw3,fl1'[aphi],
                  aphi>cw3 || aphi<=cw2,fl3'[aphi],
                  True,fl2'[aphi]];

eplot:=Plot[{gr4[aphi],gr5[aphi]},{aphi,-Pi,Pi},
           Axes->True,PlotStyle->{Hue[0.7,1,0.5]},
           Ticks->{{-Pi,{0,"0"},Pi},None},
           DisplayFunction->Identity];

mvpoint1:=Show[eplot,Graphics[{PointSize[0.04],
           Hue[0,1,1],Point[{zm,gr4[zm]}]}]];

mvpoint2:=Show[eplot,Graphics[{PointSize[0.04],
           Hue[0,1,1],Point[{zm,gr5[zm]}]}]];
(*Copy the two DO loops from program A.II.10*)
```

(*A.II.12 HELICOIDAL (SPIRAL STAIRCASE) MINIMAL SURFACE*)

(*Uses general equation for a helicoidal /catenoidal surface and sets suitable coefficients for helicoid*)

(*Requires program Parametric Plot 3D.m to be run first*)

```
a = 1.0;
b = 1.0;
co = 0.;

ParametricPlot3D[
    {a Sinh[theta] Cos[phi] - b Cosh[theta] Sin[phi],
     a Sinh[theta] Sin[phi] + b Cosh[theta] Cos[phi],
     a phi +b theta + co},
     {theta, 0, Pi, Pi/12},
     {phi, 0, 6Pi, Pi/12},

(* For view from the side use viewpoint parameters
   {2.0, -3.0, 0.0}; for view from above use
   viewpoint parameters {2.0,-3.0, 4.0}*)

ViewPoint->{2.0, -3.0, 1.0}, PlotRange->All]
```

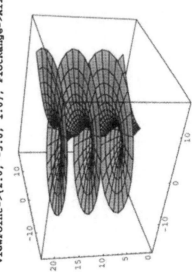

(*A.II.13 CATENOIDAL MINIMAL SURFACE*)

(*Uses general equation for a helicoidal /catenoidal surface and sets suitable coefficients for catenoid*)

(*Requires program Parametric Plot 3D.m to be run first*)

```
a = 0.0;
b = 1.0;
co = 0.;

ParametricPlot3D[
    {a Sinh[theta] Cos[phi] - b Cosh[theta] Sin[phi],
     a Sinh[theta] Sin[phi] + b Cosh[theta] Cos[phi],
     a phi +b theta + co},
     {theta, -Pi, Pi, Pi/12},
     {phi,0, 2Pi, Pi/12},

(* For view from the side use viewpoint parameters
   {2.0, -3.0, 0.0}; for view from slightly
   above use viewpoint parameters
   {2.0,-3.0, 1.5}*)

ViewPoint->{2.0, -3.0, 1.5), PlotRange->All]
```

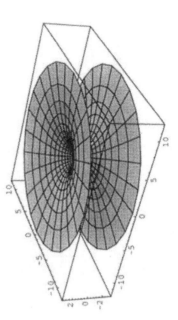

(*A.II.14 SCHERK'S MINIMAL SURFACE*)

(*Draws a restricted part of Scherk's minimal
surface suitable for visual comparison with
soap film surfaces*)

```
Scherk := Chop[Log[Cos[y]/Cos[x]], 0.00000001]
```

(*x and y must lie between -2π and 2π*)

```
Plot3D[Scherk, {x,-Pi/2, Pi/2},{y, -Pi/2,Pi/2},
    PlotRange->{-2.0, 2.0},
    PlotPoints->60,
    ClipFill->{GrayLevel[0], GrayLevel[1]}]
```

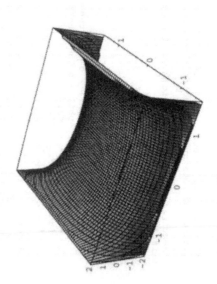

(*A.II.15 CATALAN'S MINIMAL SURFACE *)

(*Requires program Parametric Plot 3D.m to be
run first*)

```
ParametricPlot3D[
    {alpha - Sin[alpha] Cosh[beta],
     1 - Cos[alpha] Cosh[beta],
     4 Sin[alpha/2] Sinh[beta/2]},
    {alpha,-2Pi, 2Pi, Pi/6},
    {beta, -2Pi, 2Pi, Pi/6},
    ViewPoint->{-2.1, -1.1, 1.2},
    PlotRange->All]
```

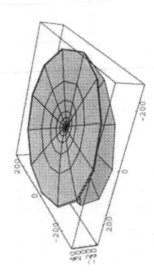

(*A.II.16(i) ENNEPER'S MINIMAL SURFACE*)

(*Requires Program Parametric Plot 3D.m to be run first*)

```
phi[f_,g_,eta_] := {(1-g^2)f, I(1+g^2)f, 2 f g} eta
X[z0_, f_, g_, eta_]:= X[z0, f, g, eta] =
    Re[ Integrate[ phi[f, g, eta], {z, z0, z}]]/.
    z->r Exp[I th]
Xeval = X[0, 1, z, 1]
ParametricPlot3D[ Xeval, {th, 0, 2 Pi, Pi/20},
    {r, 0, 2.0, 1/10),
ViewPoint ->{1.330, -1.207, 2.730), PlotRange->All]
```

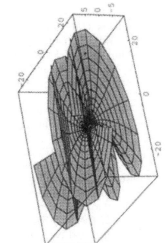

(*A.II.16(ii) ENNEPER'S MINIMAL SURFACE - ALTERNATIVE DERIVATION USING SPHERICAL COORDINATES*)

(*Requires program Parametric Plot 3D.m to be run first*)

```
ParametricPlot3D[
    {p theta - p^3 Cos[3. theta]/3 ,
    -p Sin[theta] - p^3 Sin [3. theta],
    p^2 Cos[2. theta]},
    {theta,- 2Pi, 2Pi, Pi/20},
    {p, 0, 3.0, 0.2},
ViewPoint->{8.000, -4.000, 4.000),
PlotRange->All]
```

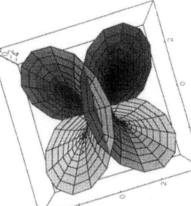

ViewPoint ->{0.050,
    -0.015, 4.000),
PlotRange->All]

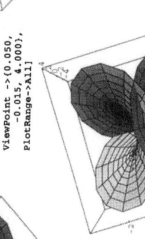

Bibliography

Adkins C J 1968 *Equilibrium Thermodynamics* (London: McGraw-Hill); 3rd edition 1983

Almgren Jr F J and Taylor J E 1976 The geometry of soap films and soap bubbles *Sci. Am.* **235** No 1 (July) 82–93

Arfken G 1966 *Mathematical Methods for Physicists* (New York: Academic)

Babcock K L and Westervelt R M 1989 Topological 'melting' of cellular domain lattices in magnetic garnet films *Phys. Rev. Lett.* **63** 175–178

Blakely J M 1973 *Introduction to the Properties of Crystal Surfaces* (Oxford: Pergamon)

Boys C V 1890 *Soap Bubbles and the Forces which Mould Them* (London: Society for the Promotion of Christian Knowledge); enlarged edition 1959 *Soap Bubbles, their Colours and the Forces which Mould Them* (New York: Dover)

Bragg L and Nye J F 1947 A dynamical model of a crystal structure *Proc. R. Soc.* A **190** 474–481

Callahan M J, Hoffman D and Hoffman J T 1988 Computer graphics tools for the study of minimal surfaces *Commun. ACM* **31** 648–661

Costa C 1984 Example of a complete minimal immersion in R3 of genus one and three embedded ends *Bull. Soc. Bras. Mater.* **15** 47–54

Courant R and Robbins H 1941 *What is Mathematics?* (London: Oxford University Press)

Cundy H M and Rollett A P 1951 *Mathematical Models* (London: Oxford University Press); 2nd edition 1961

Curl R F and Smalley R E 1991 Fullerenes *Sci. Am.* **265** No 4 (October) 32–41

Dickson S 1990 Minimal surfaces *Mathematica* **1** (1) 38–40

Du D-Z and Hwang F K 1990 The Steiner ratio of Gilbert and Pollack is true *Proc. Natl. Acad. Sci. USA* **87** 9464–9466

Erle M A, Gilette R D and Dyson D C 1970 Stability of interfaces of revolution with constant surface tension—the case of the catenoid *Chem. Eng. J.* **1** 97–109

Fejes Tóth L 1964 What the bees know and what they do not know *Bull. Am. Math. Soc.* **70** 468–481

Gaines Jr G L 1966 *Insoluble Monolayers at Liquid Gas Interfaces* (New York: Wiley)

Gilbert E N and Pollack H O 1968 Steiner minimal trees *SIAM J. Appl. Math.* **16** 1–29

Gray A 1993 *Modern Differential Geometry of Curves and Surfaces* (Boca Raton, FL: Chemical Rubber)

Greenler R 1980 *Rainbows, Halos and Glories* (Cambridge: Cambridge University Press)

Herring C 1953 *Structure and Properties of Solid Surfaces* ed R Gomer and C S Smith (Chicago, IL: University of Chicago Press)

Hildebrandt S and Tromba A 1985 *Mathematics and Optimal Form* (New York: Scientific American Books)

Hoffman D 1987 The computer-aided discovery of new embedded minimal surfaces *Math. Intelligencer* **9** 8–21

Hoffman D T 1979 Smart soap bubbles can do calculus *Math. Teacher* **72** 377–385 and 389

Isenberg C 1975 Problem solving with soap films *Phys. Educ.* **10** 452–456 and 500–503

Isenberg C 1976 The soap film: an analogue computer *Am. Sci.* **64** No 5 514–518

Isenberg C 1977 Problem solving with soap films *Phys. Teacher* **15** 39–41

Isenberg C 1978 *The Science of Soap Films and Soap Bubbles* (Clevedon, Avon, UK: Tieto); reprinted 1992 (New York: Dover)

Isenberg C 1981 Soap films and soap bubbles *Phys. Educ.* **16** 218–222

Izyumov Yu A and Syromyatnikov V N 1990 *Phase Transitions and Crystal Symmetry* (Dordrecht: Kluwer); translated from the Russian edition (1984)

Kuehner A L 1958 Long-lived bubbles; the use of sodium 9,10-dibromo-stearate solutions *J. Chem. Educ.* **35** 337–338

Landau L D and Lifshitz E M 1980 *Statistical Physics Part 1 (Course in Theoretical Physics, vol 5)*; revised and enlarged third edition by E M Lifshitz and L P Pitaevskii, translated by J B Sykes and M K Kearsley (Oxford: Pergamon). First English edition 1959

Lawrence A S C 1929 *Soap Films, a Study of Molecular Individuality* (London: Bell)

Longley W and McIntosh T J 1983 A bicontinuous tetrahedral structure in a crystalline-liquid lipid *Nature* **303** 612–614

Lovett D R 1981 Soap film analogue of Fermat's principle and Snell's law *Phys. Educ.* **16** 376–379

Lovett D R and Smith S R P 1978 Soap films in a wedge *Phys. Educ.* **13** 351–352

Lovett D R and Smith S R P 1980 Phase transitions and soap films *Sch. Sci. Rev.* **62** 287–298

Lovett D R and Tilley J 1990 A soap film model illustrating phase transitions *Eur. J. Phys.* **11** 208–214

Lovett D R and Tilley J 1991 Illustrating phase transitions with soap films *Am. J. Phys.* **59** 415–421

Mackay A L 1985a Periodic minimal surfaces *Nature* **314** 604–606

Mackay A L 1985b Periodic minimal surfaces *Physica* B **121** 300–305

Maeder R 1991 *Programming in Mathematica* 2nd edition (Redwood City, CA: Addison-Wesley)

Misner C W, Thorne K S, and Wheeler J A 1973 *Gravitation* (San Francisco, CA: Freeman)

Mitchell C E J and McLean A B 1993 Bravais lattices, surface nets and Buckminsterfullerenes *Mathematica* **3** (3) 65–68

Morgan F 1992 Minimal surfaces, crystals, shortest networks, and under-graduate research *Math. Intelligencer* **14** 37–44

Newman F H and Searle V H L 1957 *The General Properties of Matter*
    5th edition (London: Edward Arnold)
Nitsche J C C 1974 Plateau's problems and their modern ramifications
    Am. Math. Monthly **81** 945–968
Nitsche J C C 1989 *Lectures on Minimal Surfaces* vol 1 (Cambridge:
    Cambridge University Press); translation (with additions) of part of
    1975 *Vorlesungen über Minimal Fläschen* (Berlin: Springer)
Ollerenshaw K 1979a The magic of mathematics *IMA Bull.* **15** 2–12
Ollerenshaw K 1979b A note on space-filling polyhedra *IMA Bull.* **15**
    306–308
Ollerenshaw K 1980 The minimum surface connecting the sides of a
    cubic frame *IMA Bull.* **16** 54–60
Ohmae T 1993 A soap bubble model for coalescence, rearrangement,
    and splitting reactions of fullerenes *Bull. Chem. Soc. Japan* **66** 2396–
    2397
Open University 1987 *S102; A Science Foundation Course, Units 5–6,
    Into the Earth: Earthquakes, Seismology and the Earth's Magnetism*
    (Milton Keynes: Open University Press)
Osserman R 1986 *A Survey of Minimal Surfaces* (New York: Dover); an
    expanded version of the 1st edition 1969 (New York: Van Nostrand
    Reinhold)
Plateau J A F 1873 *Statique Expérimentale et Théorique des Liquides
    Soumis aux Seules Forces Moléculaires (2 vols)* (Belgium: Clemm)
Poston T and Stewart I 1978 *Catastrophe Theory and its Applications*
    (London: Pitman)
Rämme G 1992 *Soap film models Educ. Chem.* November 159–160
Roberts G 1990 *Langmuir–Blodgett Films* (New York: Plenum)
Saunders P T 1980 *An Introduction to Catastrophe Theory* (Cambridge:
    Cambridge University Press)
Schwarz H A 1890 *Gesammelte Mathematische Abhandlungen* (2 vols)
    (Berlin: Springer)
Sinclair M E 1907 On the minimum surface of revolution in the case
    of one variable end point *Ann. Math.* **8** 177–188
Stavans J and Glazier J A 1989 Soap froth revisited—dynamic scaling
    in the two-dimensional froth *Phys. Rev. Lett.* **62** 1318–1321
Stevens P S 1976 *Patterns in Nature* (Harmondsworth: Penguin); first
    published 1974
Stewart I 1991 Trees, telephones and tiles *New Scientist* **132** No 1795
    (16th November) 26–29
Stong C L 1969 How to blow soap bubbles that last for months or even
    years *Sci. Am.* **220** No 5 128–132
Taylor J E 1976 The structure of singularities in soap-bubble-like and
    soap-film-like minimal surfaces *Ann. Math.* **103** 489–539
Thompson D'Arcy Wentworth 1942 *On Growth and Form* 2nd edition
    (Cambridge: Cambridge University Press); 1st edition 1917
Von Neumann J 1952 *Metal Surfaces* (Cleveland, OH: American
    Society for Metals)
Walker J 1987a Music and ammonia vapor excite the color pattern of
    a soap film *Sci. Am.* **257** No 2 (August) 92–95
Walker J 1987b Sticky threadlike substances that tend to draw
    themselves out into bead arrays *Sci. Am.* **257** No 3 (September)
    100–103

Weaire D and Rivier N 1984 Soap, cells and statistics—random patterns in two dimensions *Contemp. Phys.* **25** 59–99

Wheeler J A 1990 *A Journey into Gravity and Spacetime* (New York: Scientific American Books)

Wolfram S 1991 *Mathematica, a System for Doing Mathematics by Computer* 2nd edition (Redwood City, CA: Addison-Wesley)

Yeretzian C, Hansen K, Diederich F and Whetten R L 1992 Coalescence of fullerenes *Nature* **359** 44–47

Zeeman E C 1972 A catastrophe machine *Towards a Theoretical Biology 4* ed C H Waddington (Edinburgh: Edinburgh University Press) pp 276–282. Also in Zeeman E C 1977 *Catastrophe Theory, Selected Papers 1972–1977* (Reading, MA: Addison-Wesley) pp 409–416

Zeeman E C 1976 Catastrophe theory *Sci. Am.* **234** No 4 65–83. Expanded version in Zeeman E C 1977 *Catastrophe Theory, Selected Papers 1972–1977* (Reading, MA: Addison-Wesley) pp 1–64

Index